Innovation in Small Professional Practices in the Built Environment

Innovation in Small Professional Practices in the Built Environment

Shu-Ling Lu
School of the Built Environment
The University of Salford

&

Martin Sexton
School of Construction Management and Engineering
University of Reading

WILEY-BLACKWELL
A John Wiley & Sons, Ltd., Publication

This edition first published 2009
© 2009 Blackwell Publishing Ltd

Blackwell Publishing was acquired by John Wiley & Sons in February 2007. Blackwell's
publishing programme has been merged with Wiley's global Scientific, Technical, and Medical
business to form Wiley-Blackwell.

Registered office
John Wiley & Sons Ltd, The Atrium, Southern Gate, Chichester, West Sussex, PO19 8SQ,
United Kingdom

Editorial offices
9600 Garsington Road, Oxford, OX4 2DQ, United Kingdom
2121 State Avenue, Ames, Iowa 50014-8300, USA

For details of our global editorial offices, for customer services and for information about how
to apply for permission to reuse the copyright material in this book please see our website at
www.wiley.com/wiley-blackwell.

Wiley also publishes its books in a variety of electronic formats. Some content that appears in
print may not be available in electronic books.

Designations used by companies to distinguish their products are often claimed as trademarks.
All brand names and product names used in this book are trade names, service marks,
trademarks or registered trademarks of their respective owners. The publisher is not associated
with any product or vendor mentioned in this book. This publication is designed to provide
accurate and authoritative information in regard to the subject matter covered. It is sold on the
understanding that the publisher is not engaged in rendering professional services. If
professional advice or other expert assistance is required, the services of a competent
professional should be sought.

Library of Congress Cataloging-in-Publication Data

Lu, Shu-Ling.
  Innovation in small professional practices in the built environment /
Shu-Ling Lu and Martin Sexton.
     p. cm. – (Innovation in the built environment)
  Includes bibliographical references and index.
  ISBN 978-1-4051-9140-1 (hardback : alk. paper)
  1. Construction industry–Research.   2. Construction industry–Technological innovations.
  3. Small business–Technological innovations.   4. Building–Technological innovations.
  I. Sexton, Martin, 1966-   II. Title.

TH213.5.L82 2009
690.068–dc22
                                                                                    2008044762

A catalogue record for this book is available from the British Library.

Set in 10/12 pt Sabon by Aptara® Inc., New Delhi, India
Printed in Singapore by Markono Print Media Pte Ltd

1 2009

Innovation in the Built Environment

Series advisors

Carolyn Hayles, *Queen's University, Belfast*
Richard Kirkham, *University of Manchester*
Andrew Knight, *Nottingham Trent University*
Stephen Pryke, *University College London*
Steve Rowlinson, *The University of Hong Kong*
Derek Thompson, *Heriot Watt University*
Sara Wilkinson, *Deakin University*

Innovation in the Built Environment (IBE) is a new book series for the construction industry published jointly by the Royal Institute of Chartered Surveyors and Wiley-Blackwell. It addresses issues of current research and practitioner relevance and takes an international perspective, drawing from research applications and case studies worldwide.

- presents the latest thinking on the processes that in fluence the design, construction and management of the built environment

- based on strong theoretical concepts and draws on both established techniques for analysing the processes that shape the built environment – and on those from other disciplines

- embrace a comparative approach, allowing best practice to be put forward

- demonstrates the contribution that effective management of built environment processes can make

Published and forthcoming books in the IBE series

Ankintoye & Beck, *Policy, Finance & Management for Public-Private Partnerships*
Pryke, *Construction Supply Chain Management: Concepts and Case Studies*
Boussabaine, *Risk Pricing Strategies for Public-Private Partnerships*
Kirkham & Boussabaine, *Whole Life-Cycle Costing*
Booth, Hammond, Lamond & Proverbs, *Solutions to Climate Change Challenges in the Built Environment*

We welcome proposals for new, high quality, research-based books which are academically rigorous and informed by the latest thinking; please contact Stephen Brown or Madeleine Metcalfe.

Stephen Brown
Head of Research
RICS
12 Great George Street
London SW1P 3AD
sbrown@rics.org

Madeleine Metcalfe
Senior Commissioning Editor
Wiley-Blackwell
9600 Garsington Road
Oxford OX4 2DQ
mmetcalfe@wiley.com

Contents

About the Authors

Dr Shu-Ling Lu, PhD, MSc, BSc, Dip (Arch)
Dr Shu-Ling Lu is a lecturer in Organisational Management of Construction within the School of the Built Environment at the University of Salford in the UK. She is the Joint Co-ordinator of the International Council for Research and Innovation in Building and Construction (CIB) Task Group 65 in the Management of Small Construction Firms. Dr Lu's main research area includes innovation management within small construction firms (particularly within knowledge-intensive professional service firms), gender issues in construction and academia–industry engagement. Dr Lu has published 2 books, 3 book chapters, and 40 journal and conference papers. Dr Lu has been invited to provide a number of keynote addresses in the areas of knowledge and quality management.

Professor Martin Sexton, BSc, MSc, PhD
Martin Sexton is a professor in Construction Management and Innovation at the University of Reading. His research interests range across the organisation and management of construction – with a particular focus on understanding the nature and process of innovation at sector, company and project levels. Martin is the Joint Co-ordinator of the International Council for Research and Innovation in Building and Construction (CIB) Working Commission 65 in the Organisation and Management of Construction. He has published widely including 2 books and over 150 journal and conference papers.

Foreword

Construction is a huge international business and 'big' is often what is reinforced with big projects and big firms receiving most of the media and press attention. However, the vast majority of construction and associated professional firms are small enterprises, employing less than 50 people delivering at least 60% of the output. It is well recognised and recorded that small to medium sized enterprises (SMEs) provide a rich source of knowledge, innovation and value-creating qualities to the economy. This is particularly true in construction where even the largest companies rely completely on supply chains made up of countless small businesses. Small professional practices must use and advance knowledge to be successful. Dr Shu-Ling Lu and Professor Martin Sexton have taken the commendable initiative to help us understand and model how these companies create, manage and exploit innovation.

Creating, maintaining and developing small professional practices are based on the notion that the business will have knowledge and expertise worthy of being sold in the market place. In other words it has knowledge capital that is valuable, but how is this capital captured, integrated, managed, exploited and developed in a business made up of highly skilled individuals and teams? To answer this question it is necessary to understand the nature of knowledge, its dimensions and variety which are critical in the creation of knowledge capital. There is benefit in understanding these issues particularly if knowledge and innovation is going to be harnessed to provide maximum return. Furthermore these are important considerations for professional firms who are attempting to create brand value. What is one of these enterprises worth if its knowledge capital is mainly held by a number of key employees who fundamentally are volunteers, who are mainly uncontrollable and are free to emigrate at any time? Unless the organisation learns from the experiences and talents its people it will not create corporate worth; it will fail to build on its existing competences and it will not deliver the benefits of teamwork. Perhaps more fundamentally its knowledge and capability will be shallow, even one person deep which can lead to professional service firms being decimated overnight by the resignation of small number of key members of staff.

It is necessary and beneficial for organisations to develop a means of exploiting not only individual knowledge, but also the combined intelligence and skills of its people. To help understand these dynamics and to test and evaluate the concepts in practice, Shu-Ling Lu and Martin have used a case study methodology to examine innovation activity in a small architectural practice drawing interesting conclusions. Not least is the importance of leadership in promoting shared values in which individual and organisational needs are addressed; where innovation is applied to solving client problems and to maintaining competitive advantage.

I have known Martin for many years and he has an excellent pedigree in construction management and research. He has developed a particular interest in small professional service firms in the construction industry working with Professor Peter Barratt at Salford University. Together they represent a tour de force in professional practice management and have provided much authoritative research into the behaviour of professional firms in the construction industry. Shu-Ling and Martin have built on their research work and that of others to produce an understandable and readable insight into innovation in small professional service firms. They have successfully unravelled the complex behavioural and organisational forces taking place and created a framework to help practitioners understand the issues and to fashion the right environment in which to foster innovation and deliver economic value. I am sure readers will find this an interesting, stimulating and beneficial experience.

Professor Trevor Mole
Managing Director, Property Tectonics
President of the European Association of Building Surveyors and
Construction Experts

Introduction

1.1　Background

The 'knowledge economy' is now significantly changing the structure of industry and the key determinants of competition. The knowledge economy is defined by DTI (1998, p. 1) as:

> ...one in which the generation and the exploitation of knowledge has come to play the predominant part in the creation of wealth. It is not simply about pushing back the frontiers of knowledge; it is also about the more effective use and exploitation of all types of knowledge in all manner of economy activity.

There is significant consensus that the knowledge economy is fundamentally based on the 'knowledge' capabilities of people (e.g. Drucker, 1997; Dougherty, 1999). It is argued that the knowledge possessed by 'staff' represent a key source of sustainable competitive advantage for individual organisations (e.g. Raich, 2002), countries (e.g. Porter, 1990; BERR and HMT, 2007) and trading blocs (e.g. EC, 2007).

The transition to knowledge economies is, to varying degrees, affecting, and being affected by, many organisations, sectors and industries. For example, evidence shows that knowledge-intensive business services account for a significant and growing proportion of economic activity in modern industrial economies (OECD, 2006; Commission of the European Communities, 2007). According to Robert Huggins Associates (2006), knowledge-based business services in 2006 account for 7.6% of the total economic output (as a percentage of total gross valued added) of the European Union (p. 1). This trend is evident in the UK. The share of knowledge-based services, for instance, in the total UK economy has risen from 5% in 1968 to 30% in 1997 (EC, 2000) and 54% of business sector value added in 1998 (DTI, 2002, p. 78). This shift towards a knowledge economy is reflected in the UK construction industry with, for example, the number of construction professional service firms rising from 19 000 in 1996 (CIC, 2003, p. 9) to 27 950 in 2005 (CIC, 2008, p. 5). Further evidence of this trend is the rise in the employment in the construction professional service firms, 'from approximately 180 000 in 1996 to 270 000 in 2005' (CIC, 2008, p. 27).

The services offered by professional service firms are characterised by being highly knowledge intensive in nature (Løwendahl, 2000). The principal means by which this growing body of professional service firms create value is through the successful creation and management of knowledge. Robertson *et al.* (2001, p. 334), for example, stress:

> Managing knowledge is a value-creating process in most organisations and is particularly important in knowledge-intensive firms.

The 'value-creating' performance of the construction industry, however, has often been questioned by its clients. The common perception of the construction industry is that of an industry which delivers products and services which are often of inappropriate quality, and which fail to meet client's demands for price certainty and guaranteed delivery. The 'Egan' report on the UK construction industry, for example, laments that 'too many of the industry's clients are dissatisfied with its overall performance' (Egan, 1998, p. 1 – emphasis added), while Fairclough (2002) has identified the need for significant performance improvement as an urgent issue.

Innovation has been described as being the principal means to bring about this improvement in the UK construction industry performance (e.g. Egan, 1998; Fairclough, 2002; Sexton and Barrett, 2003a,b; Barrett and Sexton, 2006; Brandon and Lu, 2008). The 'Egan' report recognised, for example, 'the necessary service/product improvement and company profitability can be realised through *innovations* to enhance leadership, customer focus, integrated processes and teams, quality and commitment to people' (Egan, 1998, Paragraph 17 – emphasis added). Indeed, it has been argued that '[in construction and civil engineering] innovation brings benefits of improved efficiency, effectiveness, quality of life, productivity and competitiveness' (CERF, 1997, p. 43).

Successful innovation in this book is understood to be (see Section 2.5.5 and 8.2.1):

> The effective generation and implementation of a new idea which enhances overall organisational performance, through appropriate exploitative and explorative knowledge capital which develops and integrates relationship capital, structure capital and human capital.

Small construction firms play an important part in improving the overall innovation performance of the construction industry. The growing role of small construction firms within the UK is evidenced by 99.8% of UK construction firms having less than 50 staff and employing 74.2% of the total construction workforce (BERR, 2006, Table 3). This structure is the same in the construction professional services sector, where 98% of the firms employ less than 50 people (CIC, 2008, p. i). In addition, construction projects typically draw together a significant number of diverse small and large construction firms with varying collaborations. It is acknowledged that large firms' performance is significantly affected by the performance of small firms within their supply chains (e.g. Latham, 1994; Egan, 1998). Therefore, any performance improvement of large construction firms is significantly influenced by the performance of small construction professional practices.

1.2   Research Problem

The previous section has indicated that managing knowledge is a particularly crucial issue for knowledge-intensive firms (e.g. Robertson *et al.*, 2001), and recognises that innovation is a key part in improving construction performance. There is strong consensus that managing knowledge is critical for successful innovation in small professional practices. It is argued that highly qualified knowledge workers are the core catalyst for creating and managing knowledge within such companies (e.g. Alvesson, 1995). Alvesson (1995) goes on to say that knowledge workers are engaged primarily in work of an intellectual nature. To reiterate the argument set out in Section 1.1, there is a recognition that having the right human capability within construction firms is vital to achieving successful innovation and performance improvement in the construction industry (e.g. Slaughter, 1998; Seaden *et al.*, 2001; Girmscheid and Hartmann, 2002). Within this context, the capability to innovate in small professional practices is strongly linked to the motivation and ability of the knowledge worker.

There have been a number of reports which provide guidelines to help practitioners to improve their business performance through innovation (e.g. Constructing Excellence in the Built Environment (www.constructingexcellence. org.uk)). They have provided recommendations for practices and procedures to be adopted by the construction industry and its main stakeholders to realise step improvements in both large and small construction firms. Innovation initiatives to deliver the improvements suggested in these industry guidelines, however, inadequately address project-based, service-enhanced forms of construction enterprises (e.g. Gann and Salter, 2000). Indeed, the relevance and accessibility of many of these initiatives for small construction firms are still debatable (e.g. Miozzo and Ivory, 1998; Sexton and Barrett, 2003a,b; Wharton, 2004). Egbu *et al.* (1998, p. 605) further emphasise that 'there still remains a great deal to be investigated and learned about organizational innovations within a construction environment. This is more so within the management domain of innovation where there is still a meagre amount of empirical studies that have given attention to the innovations in construction enterprises'.

There are three potential problems of this lack of explicit research into innovation in small professional practices. First, innovation theory tends to be based on manufacturing-based firms; rather than service-based firms in general, and on construction professional practices in particular (e.g. Sexton and Barrett, 2003a; Lu and Sexton, 2006). Innovation in manufacturing has been argued to be significantly different from innovation in services (e.g. Miles, 2000). For example, innovation in the manufacturing sector often emphasise research and development work leading to 'technological' novelties (e.g. Freeman, 1982; Rothwell and Zegfeld, 1982), whilst service sectors are often based on social networks leading to 'non-technical' innovations (e.g. Sundbo, 1999; Kandampully, 2002). It is this social network perspective which results in the service production process, and the final service, being more integrated, in both time and function, than in manufacturing (Sundbo,

1997), with individual innovation often consisting of process, organisation, market and product dimensions (Bilderbeek *et al.*, 1994).

Second, innovation research tends to focus on non-project-based firms in relatively stable supply chains; rather than project-based firms in relatively dynamic supply chains in general, and on construction professional practices in particular. Project-based firms are defined as those which operate on the basis of projects as their products and services need to be significantly customised to meet the particular requirements of individual clients. Projects within such firms are 'singled out as basic units, so that managerial responsibilities, resources allocation . . . and accounting data are directly or indirectly defined in terms of projects or aggregation of projects' (Warglien, 2000, p. 3). Innovation in non-project-based firms has been argued to be significantly different from innovation in project-based firms (e.g. Gann, 2000; Gann and Salter, 2000). Non-project-based firms are better able, through functional hierarchy, to own and maintain innovation compared to project-based firms. These firms engage in loose-coupled horizontal transactions between project participants and which result in project teams having fragile contexts in which to commit to, and reap reward from, innovation activity (e.g. Turner and Keegan, 1999). Indeed, Gann and Salter (2000) argue that in project-based organisation, innovation activity often relies upon resources from other companies. As a consequence of their weak appropriation of economic rent, innovation in project-based firms is seen as useful, but primarily as costly and dangerous (e.g. Keegan and Turner, 2002, Sexton and Barrett, 2005).

Finally, innovation research tends to focus on large firms; rather than small firms in general, and on construction professional practices in particular (e.g. Page *et al.*, 1999). Innovation in large firms has been indicated to be significantly different from small firms (e.g. Sexton and Barrett, 2003a,b). For example, innovation capability and outcomes of large firms tend to be more mechanistic, whilst small firms are organic in nature making them more agile and responsive (e.g. Rothwell, 1989; Nooteboom, 1994; Rothwell and Dodgson, 1994). However, small firms' innovation potential is constrained by intrinsic problems which large firms do not have. Rothwell and Zegfeld (1982) identify four challenges unique to small manufacturing firms. First, limited staff capacity and capability restrict their ability to undertake appropriate research and development. Second, small firms have scarce time and resources to allocate to external interaction. This limits the flow and amount of information on which to have discussions. Third, small firms are often affected by the excessive influence of senior management. Often small firms are vulnerable to domination by a single owner or small team who may use inappropriate strategies and skills. Fourth, small firms can have difficulty in raising finance and maintaining adequate cash flow which can result in limited scope for capital or ongoing investment in innovation activity.

In conclusion, small professional practices are becoming increasingly important agents of innovation in construction. The innovation literature, however, tends to focus on manufacturing-based, large-sized and/or non-project-based organisations. This paucity of explicit research on innovation in small professional practices ushers in real risks to policy makers, academics and industrialists of developing innovation prescriptions based on an inappropriate foundation, and thereby producing solutions for the wrong problems.

1.3   Summary and Link

This chapter has set out the background and principal focus for this book. The next chapter will contextualise the outlined research issues within the relevant general and construction-specific innovation and professional practice literature.

Key Issues from the Literature

2.1   Introduction

This chapter reviews the relevant literature which will identify and support the focal questions investigated in this research. This chapter is organised as follows:

(1)   the unique characteristics of small professional practices are discussed (Section 2.2);
(2)   the definitional debate on innovation within small professional practices is presented (Section 2.3);
(3)   the market-based and resource-based views of innovation are described (Section 2.4);
(4)   the concept of knowledge-based innovation is introduced as the principal means of achieving sustainable competitive advantage in small professional practices (Section 2.5);
(5)   the principal managerial challenges in managing knowledge capital in small construction professional practices are articulated (Section 2.6); and,
(6)   the two main questions for this research are set out (Section 2.7).

2.2   Conceptualisation of Small Professional Practices

The professional service or practice firm is the focus of a significant and growing body of relevant literature. An important starting point in this literature is the 'service' dimension of these firms. 'A service' has been usefully described as (Grönroos, 2000, p. 46):

> a process consisting of a series of more or less intangible activities that normally, but not necessarily always, take place in interactions between the customer and service employees and/or physical resources or goods and/or systems of the service provider, which are provided as solutions to customer problem.

The core of the definition above is that the generation of successful services demands a high degree of interaction and co-production of the service provision between the client and the service provider (Hansson, 2002). Extending the service concept to professional services, Hill and Neely (1988) characterise a 'professional service' as one where the client is significantly dependent on the provider to define the problem and give appropriate advice. As a consequence, professional services are associated with confidentiality, intangibility and interdependency (Glückler and Armbrüster, 2003). Such a view underlines the following remarks by Wilson (1972, p. XVI) that professional services are:

> designed to improve the purchasing organization's performance or well-being and to reduce uncertainty by the application of skills derived from a formal and recognised body of knowledge, which may be interdisciplinary, and which provides criteria for the assessment of the results of the application of the service.

The literature then moves on to argue that the principal 'provider' of these services is the professional (e.g. Maister, 1993; Løwendahl, 2000) or knowledge worker (e.g. Despres and Hiltrop, 1995). Indeed, it has been argued that the distinction between professional services and other services can be made by whether the service is done by 'professionals' or 'non-professionals' (e.g. Thomas, 1975; Kotler, 1980a; Løwendahl, 2000). Professional services are services based on the knowledge and expertise of a 'professional' (Ojasalo, 1999). A 'professional' is considered as 'someone who can act independently while bringing a body of special knowledge to bear in a work situation' (Shapero, 1985, p. 21). It is argued that professionals are highly qualified and are engaged primarily in work of an intellectual nature (Alvesson, 1995) and that professionals have a specific area of specialisation (Wheatley, 1983; Maister, 1993).

Returning to the services concept, services undertaken by professionals have been referred to as knowledge-based services (Wood, 2001). The grouping together of professionals to provide services to clients is known as a professional service or professional practice firm (Maister, 1993; Greenwood and Suddaby, 2006), a knowledge-based organisation (Winch and Schneider, 1993) and a knowledge-intensive organisation (Alvesson, 1995). The label of professional practice firms is adopted for this book (e.g. Løwendahl, 2000), as it communicates the knowledge-intensive nature of professional services and professional service firms.

To reiterate, it has been recognised that small construction firms play a dominant part in the UK construction industry (see Section 1.1). There are just under 919 000 firms with less than 50 staff in the UK construction industry (BERR, 2006: Table 3 – UK Industry Summary; Table 5 – UK Divisions). Of these, 27 950 firms are small construction professional practices (CIC, 2008, p. 5).

In summary, professional services have four principal characteristics:

(1) Professional services are knowledge-intensive in nature.
(2) Professional services are delivered by professionals/knowledge workers.
(3) Professional services are nonetheless co-produced between the knowledge worker and the client.

(4) The majority of construction professional services are provided by small firms.

Small construction professional practices thus have unique characteristics (when compared to other types of firms), and these characteristics have a significant impact on the focus and nature of innovation activity. The next section thus focuses on innovation within this context.

2.3   Definitional Debate on Innovation

There is a diverse range of definitions of innovation in the literature. Innovation is often defined as developing and implementing a new idea in an applied setting, both in the general literature (e.g. van de Ven *et al.*, 1999; DTI, 2003) and in the construction literature (e.g. Barrett and Sexton, 2006). The Department of Trade and Industry (DTI, 2003, p. 8), for example, define innovation as 'the successful exploitation of new ideas'. The 'new idea' component embraces a range of domains. Rogers (1983, p. 11 – emphasis added), for example, defines innovation as 'a product or service that is perceived as *new* by the members of the social system', and that 'it matters little whether the idea is "objectively" new as measured by the lapse of time since its first use or discovery. The perceived newness of the idea for the individual determines his or her reaction to it. If the idea seems new to the individual, it is an innovation'.

Innovation is commonly analytically separated into 'product innovation' and 'process innovation'. 'Product innovation' refers to the development and introduction of new or improved products and/or services which create or meet a new demand and which are successful in the market (e.g. Mansfield, 1991); whilst 'process innovation' involves the adoption of new or improved methods of manufacture, distribution or delivery of service which 'lower the real cost of producing outputs, although they may also give rise to changes in their nature' (Clarke, 1993, p. 143). The 'product' versus 'process' view of innovation has evolved towards a more systemic view. Athey and Schmutzler (1995) assert that process innovation (cost reducing) and product innovation (demand enhancing) are complementary. Indeed, Imai (1992, p. 226) speculates that 'process improvement and product differentiation are now being fused'. This fusion is promoting a more holistic view of innovation. The EC (1995, p. 1 – emphasis added), for example, defines innovation as:

> the renewal and enlargement of the range of products and services and the associated markets; the establishment of new methods of production, supply and distribution; and the introduction of changes in management, work organisation, and the working conditions and skills of the workforce.

This more inclusive definition is captured by the term 'organisational innovation' which is the result in the more effective use of human and physical resources; in other words, it is concerned with improving internal capabilities (Bates and Flynn, 1995).

The construction literature is generally consistent with the general litera-ture. Sexton and Barrett (2003b, p. 626) define successful innovation as 'the effective generation and implementation of a new idea, which enhances over-all organisational performance'. Similarly, the Civil Engineering Research Foundation (CERF, 2000, p. 3) defines innovation as 'the act of introducing and using new ideas, technologies, products and/or processes aimed at solving problems, viewing things differently, improving efficiency and effectiveness, or enhancing standards of living' (focusing specifically on construction profes-sional practices). Page *et al.* (1999) conclude that innovation activity within a construction context tends to gravitate around product innovation, pro-cess innovation, market innovation, organisational innovation and resource innovation.

The key common theme across the definitional debate in the literature is that 'new ideas' are taken to be the starting point for innovation. The central question which we now address is: 'What is the stimulus for these "new ideas"?' It is the investigation of this question which distinguishes the unique characteristics and challenges of innovation in small professional practices, and is the focus of the next section.

2.4   Market- and Resource-Based View of Innovation

There are two main schools of thought on the principal driver for innovation: the market-based view and the resource-based view. Each perspective will be discussed in turn.

2.4.1   Market-based view of innovation

The market-based view of innovation emphasises the role of market factors in stimulating innovation within companies. From this perspective, industry structure and the competitive environment are seen as the principal drivers of innovation (e.g. Porter, 1980, 1985). In the general literature, a number of market-based innovation theorists have investigated market or environmental influences on innovation for large firms. For example, the influences have been articulated as customer–supplier relations (von Hippel, 1988), network studies (Håkanson, 1989), market conditions (Ames and Hlavacek, 1988) and external knowledge infrastructures (Nelson, 1993). The emphasis of the market-based innovation position is that firms adapt or orientate themselves through innovation to optimally exploit changing market conditions.

The literature on market-based view of innovation for small firms, however, is rather unclear. Small firms are commonly considered downsized versions of large firms. This implies that their market-orientated innovation is based upon the market(s) they serve, and the competitive forces within that market (Porter, 1985). Storey (1994), however, finds that small manufacturing-based firms are content to survive within stable markets, often supplying one or two key customers in their local geographic market only. Their innovation strategy, therefore, is to continue with their current suppliers and customers regardless of changes in the broader market or environmental situation.

This is consistent with the view in the literature that the small professional practices' market is made up of a network of close relationships between the client and the knowledge worker. Maister (1993, p. 54 – emphasis added) asserts that '*relationships*, to remain strong, must be nurtured, and future business must be earned'. Similarly, Løwendahl (2000, p. 93 – emphasis added) stresses that 'given the high degree of independent professional judgment required in *client relations*, and the adaptation to client needs, operational authority has to be delegated to the professionals who are in direct interaction with the clients'. The principal stimulus for new ideas and thus innovation in small professional practices, it is argued, is consistent with the customer–supplier relations position advocated by von Hippel (1988). Von Hippel (1988) demonstrates that manufactures are not the sole source of innovation; rather, suppliers and customers provide a critical role. Afuah (1998, p. 72) summaries the customer as a source of innovation in the observation that 'customers who require special features in a product they use add their features to the product. If there are features that other customers can use, the manufacturer can incorporate them into its products'. The small professional practices' position, however, can be distinguished from the manufacturing perspective (where the supplier treats the clients as 'an anonymous market' to a certain extent), in that they have personalised relationships with customers who have 'a known name and face'.

The environment where this client interaction occurs is defined as 'the task environment' (Kotler, 1980b), whilst the environment where other firms compete with the firm customer and scarce resources is termed as 'the competitive environment' (Kotler, 1980b). Together the task environment and competitive environment have been labelled as 'the interaction environment' (Sexton and Barrett, 2003a, p. 36). In summary, the interaction environment is a significant market-based stimulus to innovation within small professional practices.

2.4.2   Resource-based view of innovation

In contrast, the resource-based view of innovation emphasis is that resources available to the firm, rather than on the market conditions (market-based view), are the principal catalyst for innovation (e.g. Penrose, 1959; Itami, 1987; Barney, 1991; Grant, 1995).

The resource-based view of innovation emphasis is that firms attempt to identify and nurture resources that enable firms to generate innovation to 'shape' market conditions; rather than the market-based view within advocates that market conditions 'shape' the resources which firms develop and exploit to response to opportunities and threats.

Research into small manufacturing-based firms, for example, reports that the 'accumulation and development of *resources* and *capabilities* are the relatively most important influential factors for innovativeness. Managerial skills and capabilities, internal technological resources . . . and capabilities explain to a considerable extent the differences in innovation behaviour of small firms' (Hadjimanolis, 2000, p. 278 – emphasis added). The resource-based view of innovation is evident in Wilson's (1972) argument that successful professional service firms are seen as those having the most appropriate stocks of resources for their selected innovation activities. Such a view underlines the argument by Kotler and Bloom (1984) and Løwendahl (2000) who depict distinctive competencies of small professional practices as the 'resources' and 'abilities' that a particular organisation is especially strong in relative to their competitors.

Resources in themselves are not seen as productive. Dynamic environments ceaselessly call for a new generation of resources as the context constantly shifts (Chaharbaghi and Lynch, 1999). The challenge for firms to create sustainable competitive advantage in rapidly changing and competitive environments is for resources to be integrated, coordinated and deployed as 'distinctive capabilities' (e.g. Teece *et al.*, 1997). Amit and Schoemaker (1993, p. 35) note that capabilities 'refer to firm's ability to deploy resources, usually in combination, using organizational processes, to affect a desired end. They are information-based, tangible or intangible processes that are firm specific, and are developed over time through complex interactions among the firm's resources'. Such a view underlines the following remarks by Nanda (1996, p. 97): 'while resource is a fixed asset, capability is the potential input from the resource stock to the production function'. There is agreement that capability is associated with the ability of the firm and its resources (Stalk *et al.*, 1992; Grant, 1996a).

The constant development of 'distinctive capabilities' in a dynamic environment is labelled as 'dynamic capability' (Teece *et al.*, 1997). Collis's (1994) definition of 'organisational capability' seems to have much in common with Teece *et al.*'s (1997) concept of 'dynamic capabilities' in that they both refer to the ability to develop and apply resources and skills. Collis (1994, p. 145) defines 'organisational capabilities' as 'socially complex routines that determine the efficiency with which firms physically transform inputs into outputs'. The capability of organisations to adopt, adapt and transform existing technological applications and know-how from other environments into relevant and appropriate solutions, organisational processes and technological products/services to match the sociocultural context of construction industry sector is crucial in bringing about innovation (Sexton and Barrett, 2003a,b, 2004; Harty *et al.*, 2007; Green *et al.*, 2008). The organisational capability to innovate is discussed further in Section 2.5.5.

The principal resource for small professional practices, as noted in Section 2.2, is the knowledge worker. This proposition is developed further in Section 2.5.4. In summary, it is proposed that the market- and resource-based views of innovation can be gainfully linked, by extending the argument that there is mutual adjustment between companies 'reacting to' market opportunities and threats and 'proactively' identifying, developing and exploiting resources and capabilities to secure a foundation for innovation in dynamic environments. As shown in Figure 2.1, the principal stimulus for innovation from the market-based view comes from knowledge workers' relationships with their clients, and the principal resource from the resource-based view of

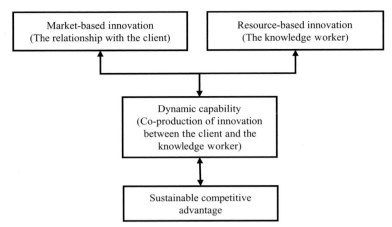

Figure 2.1   Principal sources of sustainable competitive advantage for small professional practices.

innovation is the knowledge worker. It is the proposition of this book that the development of the optimal dynamic capabilities brings these two resources together to co-produce innovation which creates sustainable competitive advantage. This view is very much an extension of similar discussions focusing on the appropriate balance between market-based and resource-based view of innovation capabilities needed in small construction firms (Sexton and Barrett, 2003a).

This section has presented the key innovation challenge facing small professional practices as the generation of an appropriate balance between market- and resource-based views. The knowledge-based view of innovation described below is presented as a way of conceptualising this balance.

2.5   Knowledge-Based View of Innovation

2.5.1   Introduction

The previous section has proposed and justified the importance of 'dynamic capability' as the driver of successful innovation and sustainable competitive advantage within small professional practices. This develops the dynamic capability concept further, and is organised as follows. First, the concept of knowledge-based view of innovation is introduced. Second, the nature of knowledge within the small professional practices is described. Third, the principal types of knowledge-based resources are identified. Finally, the main types of organisational capabilities for innovation are explored.

2.5.2   Innovation as a knowledge-based construct

It has been argued that the core capability to create and use knowledge is vital for firms to generate sustainable competitive advantage (e.g. Quinn,

1992; Drucker, 1993; Grant, 1996b; Sveiby, 1997). Leonard-Barton (1992, p. 113 – emphasis added) defines a core capability as *the knowledge set* that distinguishes and provides a competitive advantage.... A core capability is an interrelated, interdependent knowledge system'. Similarly, Peters (1994, p. 10) emphasises that 'the key source of sustainable competitive advantage is knowledge, and specifically the capacity of organisations to acquire knowledge that translates into ongoing organisational innovations'. This argument is also found within the project-based firm literature. Prenciple and Tell (2001) suggest that the ability of project-based firms to successfully innovate is determined by the knowledge they possess. Further, the theme of knowledge as a source of innovation is found within the construction literature. There is general acceptance, for example, that the management of knowledge is vital for innovation in the construction industry (e.g. Egbu *et al.*, 2000; Carrillo, 2004).

To reiterate the argument set out in Section 2.2, it has been recognised that the knowledge-intensive nature of services is the primary way to distinguish professional practices from non-professional practices, and that knowledge-based services are principally the outcome of a co-production between the knowledge worker and the client. Further, it has been emphasised that 'new ideas' are the starting point for successful innovation in small professional practices (see Section 2.3). The salient issue for small professional practices is that the 'new ideas' are intrinsically 'knowledge laden' and that they are either stimulated directly through co-production with the client or driven by contextual market needs (see Section 2.4). Muller (2001, p. 16) asserts that innovation is 'a process of knowledge creation' and that new knowledge from the process is translated into the creation of new products and services (Knapp, 1998).

The argument here is that innovation for small professional practices should be considered synonymous with a 'knowledge-based' view of innovation. Before turning to a closer examination of this view of innovation, the nature of knowledge within these firms must be understood. This is the focus of the next section.

2.5.3   The nature of knowledge within small professional practices

Knowledge has been traditionally grouped into two types: tacit and explicit (Polanyi, 1962, 1967). 'Tacit knowledge' is specific to, and resides in, individuals, and refers to knowledge that cannot be easily expressed, represented or communicated. In contrast, 'explicit knowledge' refers to knowledge which has been codified and expressed in formal language, which can be stored in databases, organisational charts, process manuals, routines and documents. The tacit and explicit distinction has evolved into knowledge as a 'noun', i.e. an 'asset' which can be neutrally articulated, stored and traded (explicit knowledge), and knowledge as a 'verb', i.e. the context-specific 'process' of knowledge creation and use (tacit knowledge). The asset and process views of knowledge, and their relevance to small professional practices, are discussed below.

An asset-orientated view of knowledge

The asset view conceptualises knowledge as 'self-contained' truths (Galliers and Newell, 2000) which can be codified and stored in knowledge repositories, and which can be shared, built upon and retained regardless of employee turnover (Wasko and Faraj, 2000). Indeed, some commentators argue that knowledge as an 'asset' forms a market, where knowledge can be traded as a 'commodity' (Davenport and Prusak, 1998). The asset view has been prevalent in the general knowledge management area (e.g. Cohen, 1998; Knock and McQueen, 1998; Alshawi, 2007) and in the construction disciplines (e.g. Egbu, 1999; Kululanga and McCaffer, 2001).

A growing body of commentators are critical of the asset view (e.g. Blackler et al., 1997), arguing that knowledge should be viewed as being relative, processional and primarily context bound (e.g. Orr, 1990; Barley, 1996). The 'process' view is the focus of the next section.

A process-orientated view of knowledge

In contrast with the asset view of knowledge, the process view of knowledge stresses the dynamic, human-centred creation and use of knowledge which is specific to a particular context and a particular time. Knowledge, from this perspective, for example, is defined as 'a dynamic human process of justifying personal belief toward the truth. Knowledge is created by the flow of information, anchored in the beliefs and commitment of its holder' (Nonaka and Takeuchi, 1995, p. 58). It follows that knowledge is 'dynamic, personal and distinctly different from data and information' (Sveiby, 1997, p. 345); and is 'information combined with experience, context, interpretation, and reflection...it is high value information that is ready to apply to decisions and actions' (Davenport et al., 1998, p. 43). There is further evidence to suggest that knowledge is a product of human reflection and experience (De Long and Fahey, 2000) and involves emotion, values and hunches (Takeuchi, 2001), and that knowledge is defined as 'a stock of expertise, not a flow of information' (Starbuck, 1992, p. 716). The common theme throughout the process view of knowledge literature is that knowledge is dynamic, humanistic and relative (Nonaka et al., 2001).

The socialisation, externalisation, internalisation and combination (SECI) model provides us with an understanding on how knowledge creation from a process view takes place between individuals, groups and organisations (see Figure 2.2). First, knowledge creation starts with 'socialisation'. The 'socialisation' mode is a process of creating knowledge by converting tacit knowledge from one entity (individual, group or organisation) to another entity. This interaction facilitates the sharing of individuals' experiences and perspectives. Second, the 'externalisation' mode is a process of creating knowledge by converting tacit knowledge into explicit knowledge. Through this process, entities articulate their formerly tacit knowledge to each other. Third, the 'combination' mode is a process of creating new explicit knowledge from existing explicit knowledge. Through this process, knowledge increasingly

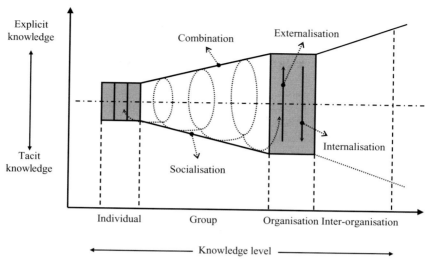

Figure 2.2 Spiral of organisational knowledge creation
(*The Knowledge-Creating Company: How Japanese Companies Create the Dynamics of Innovation* by I. Nonaka and H. Takeuchi. By permission of Oxford University Press, Inc.)

takes a concrete form. Finally, the 'internalisation' mode is a process of creating new knowledge by converting explicit knowledge into tacit knowledge. Through 'learning by doing', 'new' tacit knowledge is created, and then renews the knowledge conversion spiral. New knowledge is thus created by these four conversion processes, and through transferral of tacit/explicit knowledge from individual to group/organisational levels (Nonaka and Takeuchi, 1995).

A complementary argument is that knowledge can be categorised into individual and collective knowledge (Simon, 1957). 'Individual knowledge' is that part of the organisation's knowledge that resides in the brain and bodily skills of individuals. 'Collective knowledge' refers to the ways in which knowledge is distributed and shared among members of an organisation. Walsh and Ungson (1991) extend this argument by arguing that collective knowledge guides the behaviour, problem-solving activities and pattern of interaction between organisational members.

These two dimensions have been usefully combined to give rise to four categories of knowledge: 'embrained' (individual-explicit) knowledge depends on conceptual skills and cognitive abilities; 'embodied' (tacit-individual) knowledge is action orientated and rooted in specific physical context; 'encoded' (collective-tacit) knowledge resides in organisational routines, practices and shared norms; and 'embedded' (collective-explicit) knowledge is information conveyed by signs and symbols (Collins, 1993). Following Collins (1993), Blackler (1995) adds encultured knowledge, which is the process for achieving shared understandings and beliefs.

De Long and Fahey (2000) provide a fruitful synthesis by bringing knowledge as an 'asset' and knowledge as a 'process' dimensions together, and identify three distinct, but interactive, types of knowledge:

(1) *Human knowledge* constitutes what individuals know or know how to do, and is manifested in experience, knowledge and skills. Human knowledge is tacit knowledge.

(2) *Relationship/social knowledge* exists in relationships among individuals and groups which add value to activities. Relationship knowledge is largely tacit, composed of cultural norms that exist as a result of working together. Relationship knowledge is reflected by an ability to collaborate effectively.

(3) *Structure/structural knowledge* is embedded in organisational systems, processes, tools, rules and routines. Structure knowledge is largely explicit and rule based, and can exist independently of staff.

These three types of knowledge are proposed as being critical to understanding innovation in small professional practices. The argument here is that the appropriate generation of, and conversion between, human knowledge, relationship knowledge and structure knowledge is essential to successful knowledge creation and thus (particularly in professional practices) successful innovation. Justification for this argument is given below.

2.5.4 Knowledge-based resources for innovation

To reiterate, it is proposed that there are three types of knowledge-based resources which are critical for knowledge capital: human capital, relationship capital and structure capital. Whilst discussing these categories separately, it is important to note that there are links and synergies between each of these categories that contribute to what is being coined in this book as 'knowledge capital'. The knowledge capital is defined as 'the dynamic synthesis of both the context and process of knowledge creation and conversion between individual–organisational–individual (IOI) knowledge ba spiral, and the content of relationship capital, structure capital and human capital' and is more fully discussed in Section 2.6.

Dimension 1: Human capital

The human capital of a company is defined as 'the sum of competence, compliance and commitment' (Rabey, 2000, p. 23); and, as 'the composition of human knowledge, skills and attitude that may serve productive purposes in organizations' (Nordhaug, 1993, p. 50). These two definitions are similar in stressing that human capital represents staff motivation and ability to undertake directed and productive work. The need to create a 'high-commitment' culture of staff, in this case knowledge workers, to achieve correspondingly high levels of business performance is recognised in the human resource management literature (e.g. MacDuffie, 1995; Wood and de Menezes, 1998; The Society of British Aerospace Companies, 2002; Loosemore *et al.*, 2003).

The development and use of human capital is particularly important for small professional practices. First, knowledge workers are central to their performance. Maister (1993) indicates that knowledge workers' expertise and skills, and their ability to influence the client and perform their knowledge-intensive tasks, depend on their personal qualities (see Section 2.2). The generation of 'new ideas' (see Section 2.3) requires the motivation and in-depth knowledge and experience of knowledge workers (Baumard, 2002); thus, the capability to successfully innovate within small professional practices is significantly located within human capital.

Second, human capital is an important prerequisite condition for the 'absorption' or 'capture' of the value of knowledge into the organisational structure (see the section on structure capital below). This view is particularly important for small firms, as often a significant proportion of their knowledge about clients (relationship capital) and work activities (structure capital) are embodied in a small number of knowledge workers. This concentration of knowledge in a few staff renders small firms especially vulnerable to key people leaving the firm. As a consequence, losing key staff is potentially detrimental to small professional practices' performance (e.g. Barrett, 1993; Maister, 1993; Løwendahl, 2000). Barrett and Ostergren (1991), for instance, identify a number of adverse implications of the loss of critical staff for professional service firms, such as leaving staff taking clients with them and eroding the goodwill of the firm (see the section on relationship capital below). In summary, knowledge workers are a crucial resource in the innovation process.

Dimension 2: Relationship capital

The relationship capital has been described as social capital (e.g. Landry *et al.*, 2002), external structural capital (e.g. Sveiby, 1997), customer capital (e.g. Stewart, 1997) or relational capital (e.g. Synder and Pierce, 2002). The relationship capital is defined as 'customer and supplier relationships, knowledge of market channels and an understanding of the impact of governmental or industry association' (Bontis, 2002, p. 24); and, 'the value derived from connections outside the organization; it includes reliable suppliers and loyal customers' (Synder and Pierce, 2002, p. 478). These two definitions confirm that '[relationship] capital resides in the relationship among [human capital]' (Cohen and Prusak, 2001, p. 3). Furthermore, Cohen and Prusak (2001, p. 4) assert that '[relationship] capital consists of the stock of active connections among people: the trust, mutual understanding and shared values and behaviours that bind the members of human networks and communities and make cooperative action possible'. Social networks thus serve as a primary source of relationship capital (e.g. Coleman, 1988). This interaction develops and leverages individual's skills and knowledge (Cohen and Prusak, 2001).

The development and use of relationship capital is critical for small professional practices. In the general management literature, it has been identified that relationship capital plays a particularly important role in innovation (e.g. Ibarra, 1993; Yli-Renko *et al.*, 2001; Young *et al.*, 2001). Clients and their networks as well as the networks of the professionals are important resources for professional practices (Løwendahl, 2000). Communities of practice, for instance, have been identified as being important to the flow of knowledge within knowledge-based organisations (e.g. Hildreth and Kimble, 2004). The choice of clients influences the development of the knowledge worker (human capital), which in turn influences organisational structure (structure capital) (Scott, 1998). The importance of communities of practice has been identified in the project-based learning literature, by Ayas and Zeniuk (2001), noting that innovation is supported by reflective practitioners who share sense of purpose, a learning infrastructure and exposure to mutual role models.

Dimension 3: Structure capital

The structure capital has been described as internal structural capital (e.g. Sveiby, 1997) or organisation capital (e.g. Stewart, 1997). The structure capital has been defined as the systems for codifying, storing, transmitting and sharing knowledge (Stewart, 1997); and 'knowledge embedded in the non-human storehouses and routines of organization . . . [and] consists of mechanisms and structures of the organization that can help support employees in their quest for optimum performance' (Bontis, 2002, p. 24). Sveiby (1997) asserts that structure capital includes 'patents, concepts, models, computer and administrative systems as well as organisational culture' (p. 10).

The structure capital has been described as an important resource for small professional practices. A key aspect of the management of knowledge in organisations is the development of an organisational structure to perform knowledge-based work. Shapero (1985, p. 57) states that:

> Organisation structures and processes are concerned with configuring, channelling and affecting the ways people in the organisation relate to each other in carrying out their work.

Where knowledge is formalised and embedded in structure capital, it becomes easier (from an asset perspective) to store and to distribute to the organisation (such as by developing standardised processes, best practices, methods or organisational manuals). Information technology or information and communication technology, for example, has been recognised as an efficiency tool to improve construction industry performance (e.g. Barthorpe *et al.*, 2003). Standardisation of work (such as ISO 9000 quality management system), for instance, has been described as one way of accumulating best practices in an organisation (Thompson, 1967). As a consequence, it is believed that construction organisations should have 'a system' or 'a structure' which can support knowledge sharing interactions (Yamazaki and Ueda, 2003).

Summary

The relationship capital is the starting point for small professional practices to produce targeted services; appropriate human capital is the essential capability to bundle different resources and capabilities to form knowledge capital to bring about appropriate innovation in services and service deliveries; and, structure capital is the principal means by which outcomes of individuals' interactions can be captured, amplified and shared across different projects and across the organisation.

The key argument of this section is that knowledge capital is made up of relationship capital, structure capital and human capital. The rationale for the particular capabilities required by small professional practices to produce knowledge capital is explored in the next section.

2.5.5   Organisational capabilities for innovation

As was noted previously, innovation is produced by knowledge-based resources and capabilities (see Section 2.5.2) which form knowledge capital (see Section 2.6). There is a need to understand what kinds of capabilities are required to create, manage and exploit relationship capital, structure capital and human capital to form knowledge capital within small professional practices.

The organisational capability for innovation is defined as 'the comprehensive set of characteristics of an organization that facilitate and support innovation strategies' (Burgelman *et al.*, 1996, p. 8). It has been argued that the acquisition of 'organisational capability' may occur through the processes of 'organisational learning' (Chaston *et al.*, 1999) and that 'organisational learning' may lead to innovation (Argyris and Schön, 1996). Chaston *et al.* (1999) posit organisational learning as a necessary antecedent to building stronger core competences in organisations, particularly in small- and medium-sized enterprises. Indeed, Chaston *et al.* (2002) further indicate that the role of organisational learning in knowledge acquisition for competitive advantage is required to support the effective marketing of knowledge-based services. These viewpoints indicate the need for 'organisation learning' as a key mechanism by which firms successfully innovate.

Organisational learning can be defined as 'the process of improving actions through better knowledge and understanding' (Fiol and Lyles, 1985, p. 803), and is the 'continuous process of creating, acquiring, and transferring knowledge accompanied by a modification of behaviour to reflect new knowledge and insight, and produce a higher level assets' (Neilson, 1997, p. 2). Organisational learning is thus a process rather than an outcome (March, 1991) and results in changes in what the organisation knows and how it acts (Forss *et al.*, 1994). A key challenge for companies is when to change and when not to change. The work of March (1991) provides theoretical guidance to addressing this challenge through the distinction between exploitative and explorative routines. March (1991, p. 85) states that 'essence of exploitation is the refinement and extension of existing competencies, technologies and paradigms...[the] essence of exploration is experimentation with new alternatives'.

The term 'exploitative routines' has been described in terms of 'competitive advantage' that allows an organisation to outperform its resources in the same industry or product market. Cohen (1991, p. 136) indicates that 'improving the speed of routines and changing their detailed contents, along with the accurate switching among existing routines, are major sources of competitive advantage or other forms of organisational success'. Incremental new knowledge is thus added to the existing routines which are expected to have the end result of improving it. In other words, no attempt is made to change the paradigm, only improvements are made within the context of the prevailing paradigm. In contrast, 'explorative routines' are required to generate sustainable competitive advantage. Explorative routines consider the protection of the value of resources over time to enable the organisation to maintain its competitiveness.

It has been suggested that organisations should divide their attention and other resources between exploitation and exploration (e.g. March and Levinthal, 1999; Knott, 2002; Holmqvist, 2003; Gupta *et al.*, 2006). This view is supported by Ghemawat and Costa (1993), who argue that 'dynamic capabilities' are anchored in a firm's ability to both exploit and explore. In other words, the firm's ability to compete over time may lie in its ability to both integrate and build upon its current competencies, whilst simultaneously developing fundamentally new capabilities (Teece *et al.*, 1997).

The argument here is that there are two distinct, but interactive, types of capabilities required for successful innovation:

(1) *Exploitative capability* to utilise organisational resources to improve organisational efficiency to generate *short-term* competitive advantage.
(2) *Explorative capability* to create and use new resources and capabilities to improve organisational effectiveness to generate *sustainable* competitive advantage.

The key proposition of this section is that the concepts of exploitative and explorative capabilities are an appropriate way of understanding, connecting and managing knowledge-based resources. This proposition leads to the concept of successful knowledge-based innovation:

> The effective generation and implementation of a new idea which enhances overall organisational performance, through appropriate exploitative and explorative knowledge capital which develops and integrates relationship capital, structure capital and human capital.

2.5.6   Summary and link

This section presented 'knowledge capital' as the 'dynamic innovation capability' which generates innovation and sustainable competitive advantage within small professional practices. The literature reports that the appropriate development and use of knowledge-based resources and capabilities are critical to successful innovation, but it does not adequately address how these are developed and used in small professional practices' innovation activities. This challenge is taken up in the next section.

2.6   Key Managerial Challenges for Innovation

The co-production of professional services demands a high degree of interaction between knowledge workers and clients (see Section 2.2). Knowledge sharing and creation is thus significantly based on the human capital held by knowledge workers and others at work. Adopting De Long and Fahey's (2000) categorisation, this knowledge can be viewed as 'relationship knowledge'. Sverlinger (2000, p. 236 – emphasis added) argues that in professional service firms:

> knowledge about *market* and knowledge about *customers* [are] stored mostly in the *heads of people*.

Knowledge located within the knowledge worker can be viewed as 'human knowledge' (see Section 2.5.3). The implication of this is that relationship and human knowledge are often not effectively 'structurally' embedded within the firm; rather, they are located within the knowledge worker. This is compounded by knowledge workers tending to exhibit unique behavioural characteristics when compared to non-professionals (Maister, 1993); in particular, they are intrinsically motivated to seek challenging projects and develop new, valuable skills for themselves, i.e. their individual 'relationship knowledge' and 'human knowledge'. This individual motivation might not always be appropriately aligned to the needs of the organisation. Maister (1993), for example, states that 'brain'-type professional service organisations concentrate on complex problems which require new solutions; 'grey-hair'-type professional service organisations tend to concentrate on the firms' past experience in dealing with similar problems; and, 'procedure'-type professional service organisations usually use standard solutions to solve familiar problems. Adopting this typology, it can be argued, for instance, that for the procedural-type professional service organisation, knowledge workers who seek challenging, novel projects outside of the firm's strategic positioning can be disruptive. Similarly, for the brain-type professional service organisation, knowledge workers who focus on using 'standard solutions' will be in conflict with the firm's objectives.

Knowledge workers' understanding of customers tends to be personal and anecdotal, situationally prescribed and, according to Clippinger (1995, p. 28), 'typically neither created nor shared through traditional channels, but rather emerging and evolving from the bottom up in somewhat helter-skelter patterns'. This 'person-specific' knowledge held by knowledge workers can be labelled as 'individual knowledge' (Simon, 1957). The accrued or cumulative learning and knowledge of individuals has been referred to as 'individual knowledge capital' (Neilson, 1997, p. 1).

The challenge within small professional practices is to combine various individual knowledge domains to form dynamic 'organisational knowledge' in new configurations with feedback to, and enrich, individual knowledge. Bhatt (2002) stresses the difficulty of this challenge by stating that 'organisational knowledge' is not simply the sum of staff's 'individual knowledge'. The generation of organisational knowledge is the product of appropriate 'interaction' between individual knowledge bases (Bhatt, 2002). Organisations therefore need to develop mechanisms for tapping into the collective intelligence and skills of knowledge workers in order to create a greater 'knowledge base' (Bollinger and Smith, 2001).

The proposition made here is that organisational knowledge capital within small professional practices arises from a dynamic spiralling process wherein

Type of interaction

	Individual	Collective
Face-to-face	**Originating ba** (socialisation)	**Dialoguing ba** (externalisation)
Virtual	**Exercising ba** (internalisation)	**Systemising ba** (combination)

Media (row label at left spanning Face-to-face / Virtual)

Figure 2.3   Ba, the shared space for interaction
Reproduced by permission of SAGE Publications, London, Los Angeles, New Delhi and Singapore, from Nonaka and Teece, *Managing Industrial Knowledge: Creation, Transfer and Utilization*, 2001 (©SAGE, 2001)

relationship capital, structure capital and human capital are converted into relationship knowledge, structure knowledge and human knowledge through their exploitative and explorative capabilities. Hence, these constant interaction activities form an IOI knowledge capital spiral. Through this spiral, individual knowledge capital is converted into fresh organisational knowledge capital and allows other individuals to access the organisational knowledge capital base.

As a consequence, knowledge capital is dynamic (exploration capability), but must be capable of being accessed and used at any given time (exploitation capability). It is therefore necessary to be able to concentrate knowledge creation and conversion at a certain space and time in order to render it useful – the shared context (Nonaka and Konno, 1998). It has been argued that these 'interaction activities' take place in the 'ba' which is a place, space or facility where individuals interact to exchange ideas, share knowledge, conceptualise and create new knowledge (Nonaka *et al.*, 2001). Nonaka *et al.* (2001) differentiate four kinds of ba: (1) originating ba, (2) dialoguing ba, (3) systemising ba and (4) exercising ba (see Figure 2.3). Each ba corresponds to, and supports, a particular stage of the knowledge creation and conversion spiral.

First, 'originating ba' offers a context for the socialisation phase (see Section 2.5.3 for description of the socialisation phase). It involves sharing experiences, feelings, emotions and mental models via thought. Second, 'dialoguing ba' offers a context for the externalisation phase (see Section 2.5.3 for description of the externalisation phase). In this context, tacit knowledge becomes explicit through dialogue, reflection and the sharing mental models and skills. Third, 'systemising ba' offers a context for the combination phase (see Section 2.5.3 for description of the combination phase). Systemising ba offers a virtual collaborative environment for systemising explicit knowledge throughout the organisational structure such as databases and documentation. Finally, 'exercising ba' offers a context for the internalisation phase (see Section 2.5.3 for description of the internalisation phase). Through exercising ba, individuals learn through continuous self-refinement.

It has been argued that 'ba' may be the physical, virtual or mental ba (Nonaka *et al.*, 2001). Adopting this typology, it can be argued that 'physical ba' can be, for example, the office; 'virtual ba' could emerge from the virtual office, email, teleconferencing, telecommuting or other electronic devices; and 'mental ba' derives from shared experiences, ideas or ideals. 'Ba' provides a platform for continuously converting tacit knowledge into explicit knowledge and then back again into tacit knowledge, hence advancing collective knowledge. The various ba's provide platforms for knowledge creation and conversion to take place. The argument being made here is that the 'ba' should be focused on the 'knowledge' environment (Davenport and Prusak, 1998, p. 137). 'Ba' is thus labelled as 'knowledge ba'. For small professional practices, the 'knowledge ba' is significantly located within the interaction between individual knowledge workers and their clients. This individual level of the 'ba' can be viewed as 'individual knowledge ba'.

It has been proposed that there is a need for the shared context for knowledge creation and conversion from the 'individual level' to 'organisational level', and then back to 'individual level' (e.g. Nonaka and Takeuchi, 1995). The organisational level of the shared context can be viewed as 'organisational knowledge ba'. Organisational knowledge ba connects knowledge workers to create, share and utilise knowledge within the organisation. Knowledge at the firm level forms organisational knowledge capital.

To reiterate, individual knowledge capital within the small professional practice is mobilised and shared in the 'individual knowledge ba', where knowledge capital is held by individuals and their clients, and not necessarily held by an organisation. In contrast, organisational knowledge capital within the small professional practice is mobilised and shared in the 'organisational knowledge ba', where knowledge is held by individuals and their clients, as well as an organisation.

Organisational knowledge ba thus presents an influential factor facilitating the IOI knowledge creation and conversion spiral within small professional practices. This spiral, which continuously nurtures the interaction and development of individual and organisational knowledge ba, is taken to be the core dynamic innovation capability for small professional practices. The argument here is that knowledge capital is the dynamic synthesis of both the 'context' and 'process' of knowledge creation and conversion within 'knowledge ba', and the 'content' of relationship capital, structure capital and human capital at both individual and organisational levels.

The hypothesised 'ideal' position is thus shown on the right-hand side of Figure 2.4. The left-hand side of the diagram depicts a small professional practice where knowledge workers have very weak ties, in terms of knowledge conversion and innovation, to the 'organisational knowledge ba'. In contrast, on the right-hand side of the diagram, a stylised picture is presented of closer, more productive, alignment of individual knowledge ba and organisational knowledge ba which provides the necessary dynamic organisational knowledge capital base for successful innovation at both individual and organisational levels.

This research starts from the adaptation of the knowledge spiral model (see Figure 2.5) presented by Nonaka and Takeuchi (1995). Figure 2.5 presents 'dynamic interactions' within the small professional practice. The different levels of interaction between the firm and its client are discussed below.

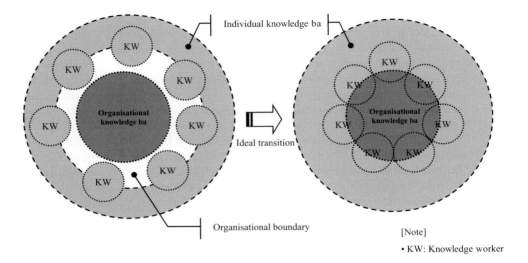

A. Weakly coupled individual and organisational knowledge capital

B. Strongly coupled individual and organisational knowledge capitals

Figure 2.4   Barriers between individual and organisational knowledge capital.

First, knowledge interactions (see Section 2.5.3 for description of four types of interaction: the socialisation, externalisation, combination and internalisation phase) start from the individual level. The interactions in a small professional practice are acquired through experience and are possessed by individual knowledge workers working with their clients. This is shown in

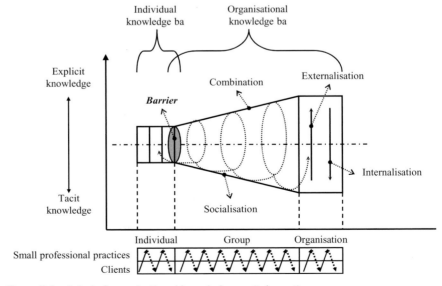

Figure 2.5   Spiral of organisational knowledge capital creation.

Figure 2.5 in the bottom rectangle. Knowledge interactions at the individual level occur in the individual knowledge ba.

Second, knowledge interactions expand outside the individual. At this stage, the collaborative interaction of individuals shares their diverse interests and issues within a team context. As the knowledge work tends to be project based (see Section 2.2), individuals are regrouped in new teams. One strategy is the development of communities of practice where groups of people deepen their knowledge through interaction on an on-going basis (e.g. Brown and Duguid, 1991). Communities of practice have potential to link individual and organisational knowledge ba's together. Knowledge interactions thus occur in the individual and organisational knowledge ba's.

Finally, knowledge interactions expand outside of the immediate team context. This implies a view of organisations as multiple communities-of-practice. At this stage, knowledge interactions occur in the organisational knowledge ba.

Figure 2.5 shows different phases of knowledge interactions between clients and the small professional practices, including individual, group and organisational interaction. It is argued that there is a paucity of research on understanding the necessary interactions between IOI knowledge ba spiral to overcome the barrier between IOI knowledge creation and conversion spiral within small professional practices. This observation may indicate that the barrier between individual knowledge ba and organisational knowledge ba is seen as the key factor which constrains the knowledge flow across individual, group and organisational levels.

The argument to this point identifies two key managerial challenges for successful innovation in small professional practices. First, they need to develop a context in which knowledge conversion takes place not only at the individual level (the knowledge worker and the client) but also at the organisational level (the knowledge worker and its organisation). Second, for this to happen, they need to motivate and equip their knowledge workers to create and engage in this context. These challenges are articulated as research questions in the next section and form the focus of the remainder of this book.

2.7   Key Research Questions

The following interconnected questions are formulated:

(1) How do small construction professional practices appropriately develop and manage knowledge interaction activities between individual–organisational–individual knowledge ba spiral, and how do these arrangements affect innovation performance?

(2) How do small construction professional practices appropriately manage and motivate their knowledge workers to create and engage in this development of, and alignment between, the individual–organisational–individual knowledge ba spiral?

2.8   Summary and Link

This chapter has provided a review and synthesis of the relevant literature pertinent to innovation in small professional practices. The central argument presented here is that knowledge-based innovation is critical for sustainable competitive advantage. It is proposed that relationship capital, structure capital and human capital knowledge-based resources and exploitative and explorative capabilities must be appropriately combined. This has led to the articulation of two research questions.

The next chapter will set out a concept of knowledge-based innovation model which will guide the investigation into these questions.

Knowledge-Based Innovation Model

3.1   Introduction

The aim of this chapter is to set out a concept model and hypotheses based on the literature synthesis set out in Chapter 2. This chapter is organised as follows. First, a concept model of knowledge-based innovation is proposed. Second, the operationalisation of the model is developed by viewing the model as a gap analysis framework. Finally, hypotheses are presented.

3.2   Description of Knowledge-Based Innovation Model

The proposed definition of knowledge-based innovation (see Section 2.5.5) forms the basis for the knowledge-based innovation concept model shown in Figure 3.1. The variables which make up the model are defined as follows:

(1) **Interaction environment** is that part of the business environment which firms can interact with, and influence, including the 'task environment' (the environment where this client interaction occurs) and the 'competitive environment' (the environment where firms compete for customers and scarce resources) (see Section 2.4).

(2) **Relationship capital (RC)** is the network resources of a firm. It results from interactions between individual, organisation, and external supplier chain partners, including reputation or image. Relationship capital is the means to leverage human capital (see Section 2.5.4).

(3) **Human capital (HC)** is defined as the capabilities and motivation of individuals within the small construction professional practices, client systems and external supply chain partners to perform productive, professional work in a wide variety of situations (see Section 2.5.4).

(4) **Structure capital (SC)** is made up of systems and processes (such as company strategies, machines, tools, work routines and administrative systems) for codifying and storing knowledge from individual, organisation and external supply chain partners (see Section 2.5.4).

(5) **Knowledge capital (KC)** is the dynamic synthesis of both the 'context' and 'process' of knowledge creation and conversion between

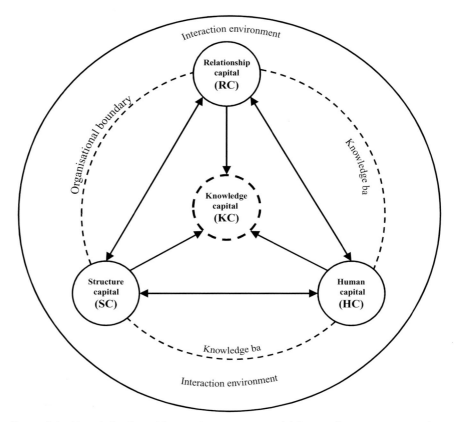

Figure 3.1 Knowledge-based innovation concept model for small construction professional practices.

individual-organisational-individual knowledge ba spiral, and the 'content' of relationship capital, structure capital and human capital (see Section 2.6).

The model proposes that interaction environment, RC, SC, HC and KC are the key variables in understanding and improving innovation performance in small construction professional practices. The variables RC, SC and HC are interdependent as indicated by the double-headed arrows. The variables RC, SC and HC contribute to KC, as indicated by the one-way arrow. All these variables need to be effectively linked for successful innovation to occur.

This conceptual knowledge-based innovation model proposes that when these variables are created and managed appropriately, they will automatically contribute to KC. From this KC, successful innovation and sustainable competitive advantage will flow. The concept model highlights the growing recognition placed by firms on the need to build, connect and energise appropriate knowledge-based resources and capabilities by providing a stimulating and supportive 'space' to generate knowledge capital from where successful innovation will spring.

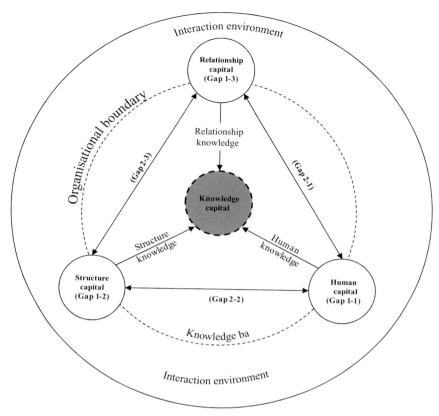

Figure 3.2   Gap analysis framework.

The operationalisation of the knowledge-based innovation model is investigated through viewing the model as a gap analysis framework (see Figure 3.2), and forms the basis for a number of indicative research questions given in Table 3.1. This gap analysis framework produces a number of hypotheses to test the research questions set out in Section 2.7. The next section will present these hypotheses.

To address the two research questions identified in Section 2.7, a meta-hypothesis and six sub-hypotheses are presented (see Figure 3.3). The general argument here is that for enduring successful innovation in small construction professional practices to take place, all hypotheses outcomes must be positive.

Table 3.1 Gaps in knowledge and understanding and their implications.

Gap		Lack of knowledge about . . .	Generic questions raised
Knowledge-based resources	1-1	Human capital	What is the human capital required for small construction professional practices for successful innovation?
	1-2	Structure capital	What is the structure capital required for small construction professional practices for successful innovation?
	1-3	Relationship capital	What is the relationship capital required for small construction professional practices for successful innovation?
Capabilities	2-1	The link between the human capital and relationship capital	How are exploitative and explorative capabilities developed and used in the interaction between human capital and relationship capital?
	2-2	The link between the structure capital and human capital	How are exploitative and explorative capabilities developed and used in the interaction between structure capital and human capital?
	2-3	The link between the relationship capital and structure capital	How are exploitative and explorative capabilities developed and used in the interaction between relationship capital and structure capital?

Meta-Hypothesis

A small construction professional practice which generates and integrates relationship capital, structure capital and human capital through exploitative and explorative capabilities will create knowledge capital for successful innovation and sustainable competitive advantage.

Knowledge-based resources

Hypothesis 1: A small construction professional practice which develops integrated individual, organisational and client human capital, structure capital and relationship capital will generate a more appropriate stock of resources for successful innovation.

Hypothesis 1-1: A small construction professional practice which develops integrated individual, organisational and client human capital will generate a more appropriate stock of human capital resources which will contribute to successful innovation.

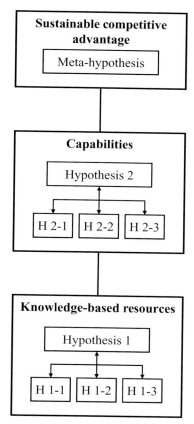

Figure 3.3   Hypotheses structure for this research.

Hypothesis 1-2: A small construction professional practice which develops integrated individual, organisational and client structure capital will generate a more appropriate stock of structure capital resources which will contribute to successful innovation.

Hypothesis 1-3: A small construction professional practice which develops integrated individual, organisational and client relationship capital will generate a more appropriate stock of relationship capital resources which will contribute to successful innovation.

Capabilities

Hypothesis 2: A small construction professional practice which generates and integrates exploitative and explorative capabilities through appropriate interaction between human capital, structure capital and relationship capital will generate appropriate knowledge capital to stimulate and support successful innovation.

Hypothesis 2-1: A small construction professional practice which generates and integrates exploitative and explorative capabilities through appropriate interaction between relationship capital and human capital will make a positive contribution to knowledge capital.

Hypothesis 2-2: A small construction professional practice which generates and integrates exploitative and explorative capabilities through appropriate interaction between structure capital and human capital will make a positive contribution to knowledge capital.

Hypothesis 2-3: A small construction professional practice which generates and integrates exploitative and explorative capabilities through appropriate interaction between relationship capital and structure capital will make a positive contribution to knowledge capital.

3.5   Summary and Link

This chapter has set out the knowledge-based innovation model which is presented as a holistic, system-orientated framework to better investigate how small construction professional practices create, manage and exploit innovation. One main hypothesis and six sub-hypotheses, based on this framework, have been articulated. The next chapter presents the research methodology used to test these hypotheses.

Case Study Methodology

4.1 Introduction

Chapter 3 set out the conceptual model and hypotheses to test the questions detailed in Section 2.7. This chapter concentrates on the design and operation of the methodology used in this research. The structure of this chapter is as follows:

(1) the overall research process within the nested methodology is introduced (Section 4.2);
(2) the case study design for this research is discussed (Section 4.3);
(3) the qualitative data collection research techniques used in this research are discussed (Section 4.4);
(4) the qualitative data analysis research techniques used in this research are presented (Section 4.5); and,
(5) the generalisability, representativeness, validity and reliability aspects of the research are set out (Section 4.6).

4.2 Overall Research Process

The overall research process used in this research is given in Figure 4.1 (based on Sexton and Barrett, 2003b, p. 624).

To reiterate, the principal interconnected challenges being investigated are as follows (see Section 2.7):

(1) How do small construction professional practices appropriately develop and manage knowledge interaction activities between individual–organisational–individual (IOI) knowledge ba spiral, and how do these arrangements affect innovation performance?
(2) How do small construction professional practices appropriately manage and motivate their knowledge workers to create and engage in this development of, and alignment between, the IOI knowledge ba spiral?

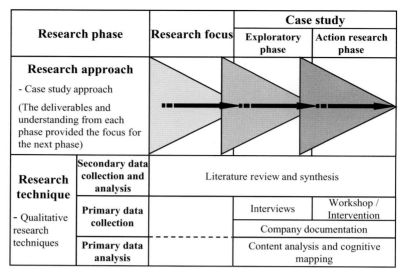

Figure 4.1 Overall research process within the nested methodology approach.

These aims were pursued through three main research phases: research focus phase, case study – exploratory phase and case study – action research phase. Each phase provided progressive focus for the next. First, the research focus phase was carried out to develop a concept model of key variables for successful innovation identified within the literature (see Section 3.2). Second, the exploratory phase of the case study was carried out to test these variables by investigating successful/unsuccessful innovation within the case study company. Finally, in the action research phase, the key findings from the exploratory phase were fed into a company workshop. The results of the exploratory phase were reviewed in the workshop by senior management of the case study company, and a high priority business improvement need identified. This need formed the basis of the intervention in the action research phase. This action research phase further tested and developed the concept model.

4.3   Case Study Design

This section examines the case study design and describes and justifies the following elements: the unit of analysis, the sampling strategy for case study firm selection, the sampling strategy for interviewee selection and the data collection techniques.

4.3.1   Unit of analysis

The definition of 'the unit of analysis' is a 'phenomenon of some sort occurring in a bounded context' (Miles and Huberman, 1994, p. 25) and

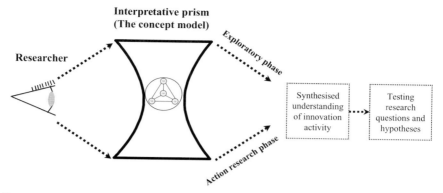

Figure 4.2   The unit of analysis within this research.

should be 'related to the way the initial research questions have been defined' (Yin, 1994, p. 22). An appropriate unit of analysis is critical, as it influences the subsequent lines of inquiry within a case study. The unit of analysis is taken as the 'innovation activity' (see Figure 4.2), i.e. the generation and implementation of an innovation is investigated through the 'interpretative' prism of the organisational model of innovation (see Section 3.2). In the exploratory phase, seven innovations were identified for investigation (see Section 5.4); whilst in the action research phase, the unit of analysis was the interim project review process innovation (see Section 6.2.1). The individual innovation activity from the exploratory phase and the action research phase helped to gather a synthesised understanding of organisational innovation activity. This synthetic understanding from the exploratory phase and the action research phase was used to investigate the research questions (see Section 2.7) and test the hypotheses (see Section 3.4).

4.3.2   Strategy for sampling design

Sample size

A longitudinal single case study of 22 months was the basis for this research. A single case has been described as the opportunity to study several contexts within the case, a number of different cases in the single firm, or the number of cases studied can be different from the number of firms (Mukherjee *et al.*, 2000). The situation where a single case study is appropriate has been argued as follows (Yin, 1994, pp. 38–40):

(1)   where the case presents a critical setting for testing an existing theory, whether the goal is to confirm, challenge or extend it;
(2)   where the case has unique or extreme characteristics; or,
(3)   where the case study exists in a situation whereby an investigator has an opportunity to observe and analyse a phenomenon previously inaccessible to scientific investigation.

The single case study adopted in this research is principally stimulated by the first and third situations above. It is believed that the single case study is best suited to dealing in an in-depth way with the multitude of fragmented perspectives and complexity of organisational life within small construction professional practices (rationale 3 above) that have been identified as important issues in Chapter 2 (rationale 1 above). Effort has been made to select a representative small construction professional practice (see below); therefore, rationale 2 above is not applicable.

The single case approach, however, has a number of limitations. The first limitation is the degree of generalisablitiy of the conclusions, models or theory development from one case study. Second, the results from a single case study can be inappropriately integrated. Leonard-Barton (1990), for example, argues that these include the risks of misjudging the relevance and impact of a single event, and of exaggerating easily available data. This research adopts the position set out by Yin (2003, p. 39) in that the results are generalised to theory (which is analogous to the way in which scientists generalise from experiments to theory) rather than to the wider population of small construction professional practices.

The risk to the generalisation to theory have been reduced in this research by focusing on a longitudinal, 22-month case study which offers a richer 'dynamic' picture than offered by, arguably, that 'snapshot' insight gained from a number of short case studies (see Section 4.6). Further, triangulation was employed to ensure robustness of data collection and analysis (see Sections 4.4 and 4.5). This view is supported by Stake (1994, p. 242), who states: 'generalization from differences between any two cases are much less to be trusted than generalizations from one'.

Sample representativeness (selection criteria)

Selection criteria for the single case study company were made on the basis of the size and type of organisation. Each criterion is discussed below.

Size of organisations

The research focuses on small firms (see Section 1.2). According to the BERR (2006), small companies are defined as those having less than 50 staff. The single case study company (see Section 5.2), labelled hereafter as ArchSME for confidentiality reasons, meets this criterion by having 40 staff.

Classification of organisations

The research focuses on construction professional practices. Adopting the definition of this firm type (see Section 2.2), it can be argued that architectural, building services, building surveying and quantity-surveying firms are examples of construction professional practices (e.g. CIC, 2008, p. 3).

The case study firm was an architectural practice. There are two reasons for this choice. First, there is evidence that 'the architectural service' is the 'archetype' of a professional practice, being almost entirely reliant on the knowledge and expertise of individual organisational members (e.g. Day and Barksdale, 1992; Boström, 1995; Wilson, 1997; Woiceshyn and Falkenberg, 2008). Second, the important role of architects within UK construction professional practices is demonstrated by architecture accounting for 41% of total fee income in the construction professional services sector (CIC, 2008, p. 6, Table 3.1).

The case study firm is a recent start-up with three equity directors within a limited liability partnership structure. The firm has grown from an initial turnover of £0.3 million in 1999 to £2 million in 2004. In the same period, the firm has increased its staff numbers from 12 to 40. The firm is at an early stage in its development and, drawing from Churchill and Lewis (1983) stage theory, is very much at the take-off and success phase. It is anticipated that the research results reported here would vary if the case study firm had been at a different phase, as the approach to innovation management evolves in unison with the developmental stage of the firm (Sexton and Barrett, 2003b).

4.3.3   Sampling strategy for interviews

Before conducting interviews, the appropriate size and composition of the interviewee sample was determined. This view is in alignment with Leedy (1988, p. 158) who argues that 'no matter how good the gathering of data is . . . the survey cannot be accurate if the people in the sample are improperly selected'.

Professional practices usually structure their employees in three levels: junior, manager and senior (Maister, 1993, p. 4). The different levels are characterised by staff with particular experience and skills sets undertaking different types of work. Senior-level professionals and middle-level professionals (managers) are highly experienced and skilled. It is argued that senior management are engaged, to a significant extent, in organisational management activities, whilst managers focus on project management activities. Managers are often project management professionals (Maister, 1993, p. 5). Junior-level professionals are primarily engaged in undertaking project tasks under the direction of the project manager. Figure 4.3 shows the structure of ArchSME using this classification. The sample set of five interviewees in the exploratory phase represents all three levels. This reduced the risk of the results being biased to a particular professional level within the firm.

Table 4.1 summarises key features of the five respondents which participated in the exploratory phase in this research. It can be seen that the average age of the respondents is fairly young (29 years), respondents have explicit architect education and qualification and that four out of five respondents come from small- to medium-sized, private architectural or building firms.

Table 4.1 Profile of respondents in interviews.

Respondent	Classification	Age	Formal qualification (graduate and fully qualified members of professional institutions)	Number of years with ArchSME and role	Number of years and role with previous firm	Previous employer		
						Main products/services	Type	Firm size
A	Senior	34	Architecture diploma Royal Institute of British Architects	2/Associate director	3/Architect 4/Architect 2/Architect	Architectural practice Architectural practice Architectural practice	Private Private Private	Medium Micro Small
B	Junior	26	Trained to Higher National Certificate in Architecture	2/Architectural technician	5.5/Technical drawing	Architectural practice: design scheme for the building, achieve partnering information and help the team building	Private	Micro
C	Manager	28	Architecture diploma Royal Institute of British Architects	3.5/Architect	5/Managing contracts on site 2/Architect	Building company Architectural practice	Private Private	Small Small
D	Manager	31	Architecture diploma Royal Institute of British Architects	3/Project architect and team leader	3/Training architects	Architectural practice	Private	Medium
E	Senior	26	Architecture diploma Masters in business administration	5/Development manager and architectural assistant	5/Estate agent 1/Copy typist 5/Selling shoes	Selling houses Preparing documents for court Children shoe shop	Private Public Private	Small Large Large

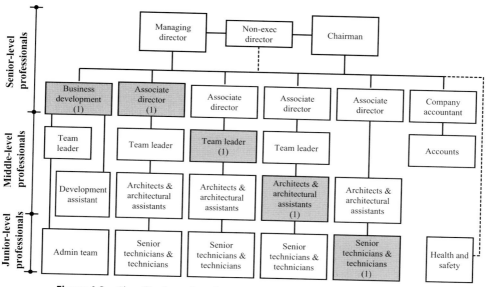

Figure 4.3   Classification of professionals within ArchSME.

4.3.4   Case study data collection design

The overall activity in the 22-month case study is given in Figure 4.4. The case study started in April 2003 and ended in January 2005. There were two main phases in this study: the exploratory phase and the action research phase. Each phase is discussed below.

Exploratory phase

The exploratory phase was 12 months in duration. The main activities within the exploratory phase are listed in Table 4.2 (see Chapter 5 for the description of the exploratory phase).

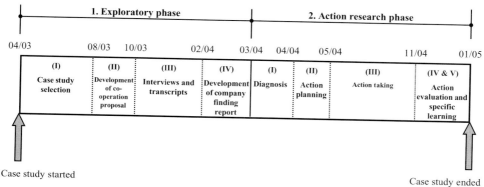

Figure 4.4   Case study phases and activities.

Table 4.2   Exploratory phase activities (April 2003 to May 2004).

Phase		Duration	Case study research activity	Outcome
I	Case study selection	April to July 2003	Emailed and telephoned around 300 small construction professional practices with research proposal	ArchSME selected
II	Development of cooperation proposal	August to September 2003	Developed cooperation proposal with ArchSME senior management Access to company documents (see Appendix A)	A confirmation email from senior management An agreed detailed company cooperation proposal An agreed revised detailed company cooperation proposal (see Appendix B)
III	Interviews and transcripts	October 2003 to January 2004	Arranged the interview schedule with ArchSME senior management Emailed interview cooperation proposal (see Appendix C) and interview protocol to each respondent (see Appendix D)	A confirmation email from senior management A confirmation email from each respondent
			Face to face interviews with each respondent Each interview was appropriately 1.5 hours duration Check transcription accuracy with each respondent	Delivered transcripts/the word-processed documents to each respondent A confirmed transcription accuracy email from each respondent
IV	Development of company finding report	February to March 2004	Developed company finding report with ArchSME senior management	A general company finding report (see Appendix E)

Action research phase

This research adopted an action research methodology (see Section 4.5), adopting the five-step process of diagnosis, action planning, action taking, action evaluation and specifying learning (Susman, 1983) (see Figure 4.5). The focus of each phase was tailored to meet the specific nature of this study, and is set out below.

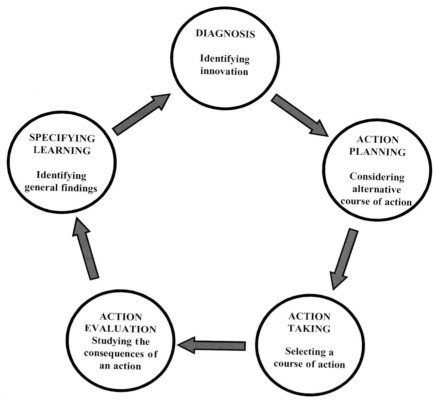

Figure 4.5   The process of action research.

Diagnosis phase

The diagnosis phase generally corresponds to the identification of the issue (be it an opportunity or problem). In this research, the 'issue' is innovation activity, and the diagnosis phase concentrated on collecting and analysing relevant information to develop a clear understanding of relevant factors.

Action planning

The action planning activity specifies organisational actions to progress the intervention, for example, milestones and deliverables.

Action taking

Action taking is to implement the action plan. The intervention within this research was carried out in six activity episodes.

Action evaluation

After the actions are completed, the action evaluation activity takes place to determine whether or not the implemented innovation has been a success or a failure as defined by the performance criteria set out in the action planning phase.

Specifying learning

Specifying learning is to reflect on the knowledge gained in the action research. The learning feeds into the knowledge capital of both the case study firm and the action researcher.

The five phases within overall action research process do not take place in five, sequential phases; rather, mini cycles, from diagnosis through to specifying learning, took place through out the action research process (see Figure 4.6). The important characteristic of each cycle is that diagnosis before

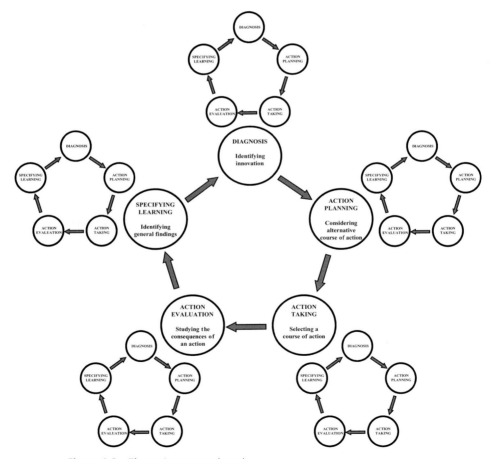

Figure 4.6   The action research cycle.

action planning, action planning before action taking, action taking before action evaluation and reflections on specifying learning. The specifying learning at the end of each cycle feeds into the diagnosis for the next cycle.

The action research phase was 10 months in duration. The main activities within the action research phase are listed in Table 4.3 (see Chapter 6 for the description of the action research phase).

4.4   Research Techniques: Qualitative Data Collection Techniques

The data collection techniques for this research consisted of reviewing relevant literature and company documentation, carrying out interviews and setting up and attending workshops and meetings. Each technique is discussed below.

4.4.1   Literature review

It is believed that prior theory in the area of research interest in the case study research should be identified through a literature review (e.g. Miles and Huberman, 1994; Yin, 2003). The literature review embraced two main areas, with a particular focus on small construction professional practices: the management of knowledge and the management of innovation. This research adopted the hermeneutic-based philosophy of interpretation of pre-understanding/understanding (e.g. Baleicher, 1980). Figure 4.7 shows the process of literature review and synthesis used in this research. Three generic strands ran through this process. The pre-understanding of the researcher represented the researcher's prior knowledge, insights and experience which the researcher drew upon to interpret a piece of general management literature. The understanding gained provided an appropriate focus to scrutinise a piece of construction-specific literature. This shaped the next phase of pre-understanding used to interpret a second piece of general management literature, and so on. The ongoing review and synthesis of the relevant literature resulted, initially, in the formulation of the research questions, and then supported the data collection and analysis activity.

4.4.2   Interviews

The interview technique is a flexible and commonly used research tool (Breakwell, 1995) and particularly appropriate if sensitive or complex questions need to be asked (Hussey and Hussey, 1997). The interview technique used in this research aimed to gain an insight into the 'below the surface activities' (Oppenheim, 1992) in terms of obtaining an overall picture of the case study company and its innovation activities.

Traditionally, there are three broad types of interview: structured, unstructured and semi-structured (e.g. Fontana and Frey, 2000). A structured interview is where a fixed schedule of questions is followed with each respondent. An unstructured interview is where the process can be shaped to the

Table 4.3   Action research phase activities (April 2004 to January 2005).

	Phase	Duration	Action research activity	Outcome
I	Diagnosis	April 2004	Presented the key findings from the exploratory phase in the company workshop	Discussed and validated the analysis and results
			Possible interventions identified and discussed	Interim project review process innovation identified
			Emailed company workshop minutes to ArchSME senior management (see Table 6.1)	A confirmation email from senior management
II	Action planning	May 2004	Developed an action plan (see Table 6.2)	A confirmation email from senior management
III	Action taking	May to November 2004	Developed first draft of the interim project review policy, guidelines and checklists Reviewed relevant company documents (see Appendix A)	The first draft of the interim project review process
			Emailed the first draft of the interim project review process to ArchSME's quality representative Meeting with ArchSME's quality representative	The third version of the interim project review process
			Emailed the third version of the interim project review process to ArchSME's quality representative Interim project review handbook reviewed by ArchSME management board	ArchSME's senior management comments on the third version of the interim project review process
			Emailed the revised interim project review handbook to ArchSME's quality representative Meeting with ArchSME senior management	QW01 ArchSME guidelines for interim project review
			Interim project review procedure reviewed by ArchSME's external ISO consultant Meeting with ArchSME senior management	QW1 interim project review handbook (Revision A)
			Emailed the revised QW1 interim project review handbook to ArchSME's quality representative Meeting with ArchSME's quality representative	QW1 interim project review handbook (Revision B)
IV and V	Action evaluation and Specifying learning	December 2004 to January 2005	Tested the interim project review process	By the end of December 2004, the interim project review process had not been implemented

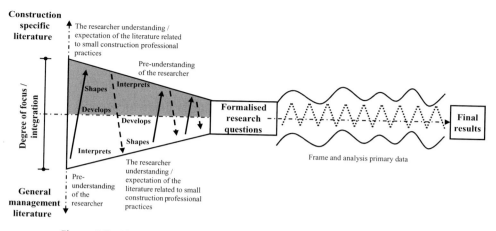

Figure 4.7   Literature review and synthesis process.

individual situation and context. There are no fixed questions, although there is often a 'checklist' of issues to be explored. The interview is conducted in an open-ended way to allow the discussion to evolve in an organic fashion. A semi-structured interview is where guidance is given in an informal setting and where a broad formalised questions are asked. The key distinction between an unstructured interview and a semi-structured interview is whether or not the interventions are made by the researcher (e.g. Royer and Zarlowski, 2001, pp. 147–148). In an unstructured interview, the researcher makes no intervention to direct the subject's remarks. This research investigated the case study company's innovation activity, with respect to a particular set of propositions set out in the concept model, research questions and hypotheses. A level of intervention by the researcher was thus required to ensure that these prepositions were adequately investigated. An unstructured interview, therefore, was not appropriate for this research. In summary, this study used a semi-structured interview during the exploratory phase.

Before starting the interviews, a semi-structured interview protocol was prepared and pretested (see Appendix D). First, the focus and content of the interviews were co-developed with a senior member of the firm – the securing of buy-in and shared ownership of the interview process by the owners of the firm were essential to the freeing up of staff time to undertake the interviews. The questions within this protocol were designed to investigate the definition of knowledge and innovation (see Section 2.5.5) and the knowledge-based innovation conceptual model (see Section 3.2). The semi-structured interview protocol (see Appendix D) was structured into four main sections: introduction, background, knowledge-based innovation details and additional information. Each section is described below.

The *introduction* section was designed to introduce this research and the researcher to the respondents.

The *background* section was designed to understand the background information of the respondent, the case study company and the company's principal clients. It helped the researcher to understand the company's business environment, its major clients and the qualifications and backgrounds of its staff.

The *knowledge-based innovation details* section was designed to understand the nature of innovation activities in ArchSME and to identify the type of resources and capabilities used. The questions in this section investigate the six variables of the knowledge-based innovation conceptual model (see Section 3.2): interaction environment, K ba, RC, SC, HC and KC. There are four subsections under this section.

The first subsection had two opening questions which were designed to understand what respondents understood by the terms knowledge and innovation. The second subsection focused on developing questions to understand the interaction environment of the company, including company business strategy, innovation strategy and the company-supporting innovation activities (RC, SC, HC, KC and K ba). The third and fourth subsections investigated successful and unsuccessful firm-specific innovation generated over the last 2 years. The identified innovations were explored by understanding how the company generated and implemented new ideas, and identifying innovation performance measurement/indicators.

The *additional information* section was designed to capture issues considered relevant by the respondents which were not raised in the interview.

A director identified key respondents at senior, middle and junior levels within the firm. When agreement to cooperate was received, the interview cooperation proposal (see Appendix C) and the semi-structured interview protocol (see Appendix D) were sent to these respondents prior to the interview. This was to allow them to know the type of issues that were going to be discussed and, where necessary, gather relevant information beforehand.

Each interview was between 1 and 2 hours in length, and was carried out face-to-face. The interviews were recorded. Prior permission to record the interview was secured from the respondents. In addition, the researcher took notes, although this was kept to a minimum to avoid unnecessary disruption during the interview. The combination of using an audio recorder and making notes is recommended for conducting interviews (e.g. Hussey and Hussey, 1997). The interview recordings were transcribed verbatim. The transcripts were sent to each participant to check for accuracy before being analysed.

4.4.3   Company documentation

In addition to the interviews, further data were obtained through the analysis of company documents. However, it was found that there was little ArchSME company documentation. This indicates, in itself, the very limited extent to which information is codified in small firms. The dearth of documentation, from a research methodology perspective, reduces the scope to triangulate participant accounts against secondary sources (e.g. Guran and Blackburn, 2001; Lu and Sexton, 2004). Appendix A gives a list of the company documentation examined.

4.4.4   Company workshop

The workshop was undertaken in April 2004 at the start of the action research phase of the case study. The workshop began with a presentation of the key findings from the exploratory phase. The remainder of the workshop was structured around a number of main questions given in the company finding report (see Appendix E), namely: what is the current position? what are the potential problems? why manage knowledge? what are potential improvement areas to sustain current growth? and what are the immediate innovations which the firm should progress? The company report was co-authored by the researcher and ArchSME's senior management. This helped to ensure the report was appropriate in focus and style, and assisted in creating shared ownership of the report, and the subsequent action research phase.

The workshop participants debated the immediate potential innovations identified in the company general finding report – beginning with exit interview process and post-project review process – followed by a discussion of how these two themes could be developed. The senior manager identified that the development and implementation of that interim project review process was needed and should be the focus of the action research phase.

The workshop was videotaped, with prior permission of the participants, for subsequent analysis. In addition, in order to maximise consensus and the commitment of the participants, the minutes of the workshop were sent to workshop members for confirmation that the discussion had been interpreted correctly (see Table 6.1).

4.5   Research Techniques: Qualitative Data Analysis Techniques

The primary data collected in this research was qualitative in nature (see Section 4.4). Content analysis and cognitive mapping data analysis techniques were used. The justification for using these techniques is twofold.

First, the content analysis technique enabled the identification of key issues from the large volume of interview transcripts (e.g. Weber, 1985). Second, in order to help the researcher to see the relationships between different ideas and perspectives emerging from the content analysis, the cognitive mapping technique was used. It is argued that the cognitive mapping technique allows the key concepts and relationships articulated by the researcher to be externalised and synthesised in a clear layout that facilitates critical enquiry and reflection (e.g. Eden, 1992). The combination of content analysis and cognitive mapping is supported by Allard-Poesi et al. (2001) who stress that the two techniques are commonly and appropriately used in management research.

The data analysis used two software packages – 'QSR NUD*IST Vivo' (NVivo), a content analysis tool (www.qsrinternational.com); and, 'Decision Explorer', a cognitive mapping tool (www.banxia.com). In the exploratory phase, the data for each identified innovation (see Section 5.4) was analysed using content analysis to develop the keynotes or variables (the presence of certain words or concepts within texts or sets of texts). These notes are in

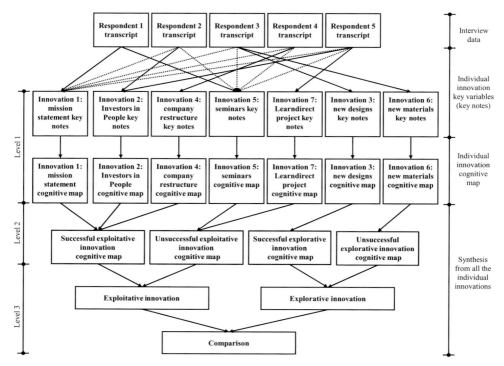

Figure 4.8   Primary data analysis structure.

short phrases (see Section 4.5.1). These notes then were fed into the cognitive mapping. Figure 4.8 shows the journey that being made by the researcher in conducting the primary data analysis.

Primary data from the five respondent transcripts were imported into NVivo's database. Three levels of analysis were articulated to identify patterns within the data. The first level consisted of the analysis of the individual innovations. First, appropriate variables (notes) were identified by the researcher's interpretation using NVivo. Second, the interrelationships between these variables were identified by the researcher's interpretation using Decision Explorer.

The second level consisted of a cross-innovation analysis and then the grouping of innovations with similar patterns. First, seven innovations were grouped into the matrix of successful/unsuccessful and explorative/exploitative innovations in order to focused insight from the data (see Section 5.4). Four types of innovations – successful explorative innovation, unsuccessful explorative innovation, successful exploitative innovation and unsuccessful exploitative innovation – were identified. Second, the interrelationships between variables of these four types of innovations were identified by the researcher's interpretation using Decision Explorer.

The third level was a summation of all the innovations within the knowledge-based innovation concept model. First, four types of innovations were grouped into two types of innovation: explorative innovation and exploitative innovation. Second, the comparison between them was made.

A noted system used in this research comprised 'free nodes', 'tree nodes' and 'sets'. The note in 'free notes' presented as unorganised or not belonged in hierarchies of categories and subcategories. The notes in 'tree notes' were presented in hierarchies of categories and subcategories. A 'set' is a grouping of nodes for purpose of working with them together. Figure 4.9 shows the structures of the note system used in this research.

The first level categories of 'tree notes' used in this research were: why mission statement successful (innovation 1), why Investors in People (IiP) successful (innovation 2), why new designs successful (innovation 3), why company restructure successful (innovation 4), why seminars failed (innovation 5), why new materials failed (innovation 6) and why Learndirect project failed (innovation 7) (see Section 5.4). The second level subcategories of 'tree notes' were critical variables for the identified innovation.

For further analysis, the nodes were also managed in sets. The first level categories of 'sets' used in this research were: successful explorative innovation (innovation 3), unsuccessful explorative innovation (innovation 6), successful exploitative innovation (the combination of innovation 1, 2 and 4) and unsuccessful exploitative innovation (the combination of innovation 5 and 7) (see Section 5.4 for the description of company innovations). The second level subcategories of 'sets' identified the critical variables for each innovation.

The process developed is illustrated using innovation 1 (the ArchSME mission statement) as an example (see Figure 4.10).

4.5.1   Content analysis

Transcripts from the five respondents were transferred into a text file in order to import it into the NVivo system. The researcher then 'interpreted' the text into 'notes' (or variables). To identify and bring together the data passages that seem to belong at a category is called coding (Richards, 1999, p. 55). Each note was coded under subcategories of 'why mission statement successful' (innovation 1). Take number 2 note: 'chairman driven' as an example, Figure 4.11 shows the context of passages coded under this category. Similar notes were combined and structured into new categories. When subcategories grew too big, they were broken down into new subcategories. The ongoing process resulted in the formulation of appropriate notes.

The research results indicate that there are 28 notes (variables) within innovation 1: mission statement (see Figures 4.12 and 4.13). It shows that 'informal presentation/workshop' (number 19 note) (see Figure 4.13) was the key element in enabling this innovation success which was referred to 17 times within the transcripts. The second highest criteria to enable this innovation success were 'informal discussion in the office' (number 4 note) (see Figure 4.12) and 'no specific way to measure the performance' (number 21 note) (see Figure 4.13). They both were referred to 13 times.

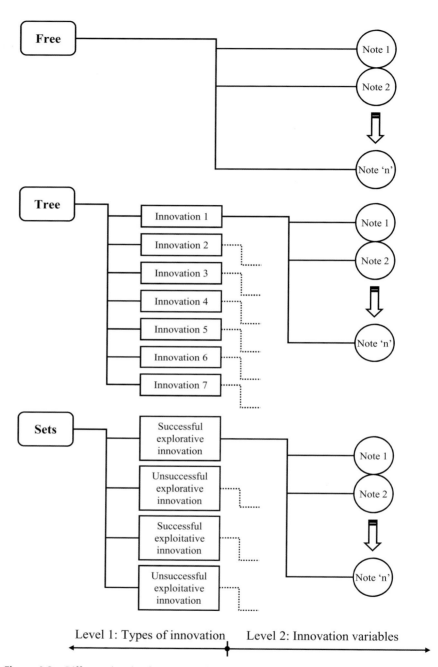

Figure 4.9   Different levels of notes used in this research.

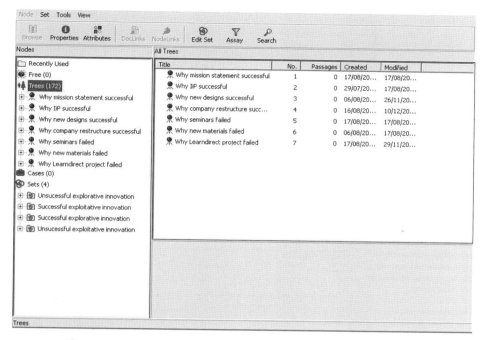

Figure 4.10   Different levels of notes produced in NVivo.

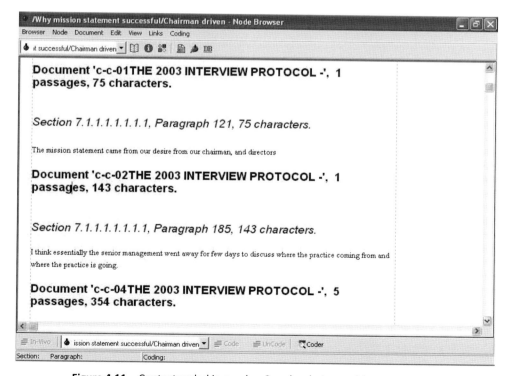

Figure 4.11   Context coded in number 2 node: chairman driven.

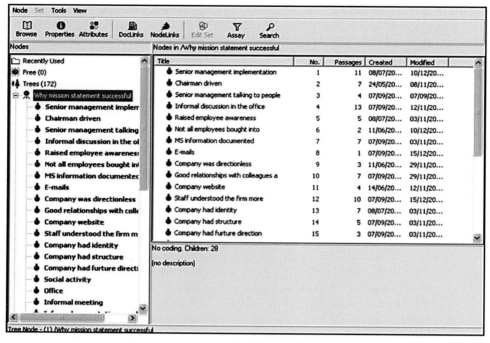

Figure 4.12   An example of key notes produced in NVivo (1).

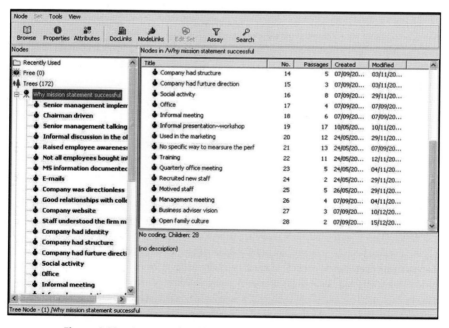

Figure 4.13   An example of key notes produced in NVivo (2).

The next section discusses how the key variables identified in NVivo were imported into Decision Explorer's database and how interrelationships between these variables were identified.

4.5.2   Cognitive mapping

In order to analyse the interrelationships between the 28 keynotes, the cognitive mapping technique was used. Two processes were conducted in order to transfer the file in NVivo's database into Decision Explorer's database. First, the keynotes coded in innovation 1 mission statement were exported as a 'NUD*IST' file. Second, this file was imported into Decision Explorer's database. Through this process, the '28 notes' under innovation 1 category and 'one' innovation 1 category coded in NVivo became '29 concepts' in the Decision Explorer system (number 1 to number 29 concepts) and produced a basic map.

In order to more easily interpret and identify the interrelationships between the 29 concepts, the four variables identified in the knowledge-based innovation conceptual model were used to form subcategories: human capital, structure capital, relationship capital and knowledge ba (see Section 3.2). In addition, in order to understand the outcome of innovation 1, one subcategory – impacts from it – was added. The total number of concepts, therefore, increased from 29 to 34.

Links were used to identify the meaning between variables. A link is represented as an arrow. In this research, an arrow represented the phrase 'leads to' or 'cause'. Taking number 16, 18 and 10 concepts as an example (see Figure 4.14), number 16 (1 16) 'social activity' and number 18 (1 18) 'informal meeting' have implications for, or lead to number 10 (1 10) 'good relationships with colleagues and suppliers'.

Figure 4.15 shows a cognitive map of innovation 1 'why mission statement successful' created in Decision Explorer.

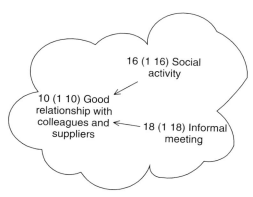

Figure 4.14   An example of linking concepts.

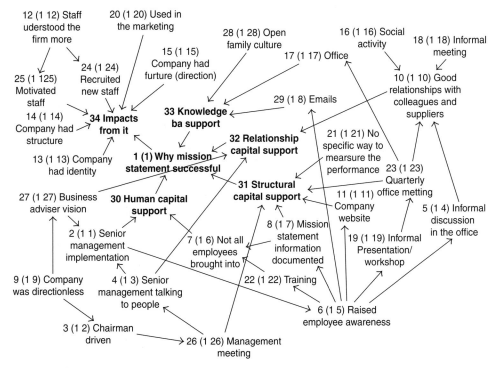

Figure 4.15   An example of innovation 1 cognitive map produced in Decision Explorer.

The data analysis rationales and procedures have been identified and discussed. The following section discusses the procedures followed to ensure the validation of the research methodology.

4.6   Validation – Triangulation Strategy

This section examines the validation of the results from the ArchSME case study. Different research approaches and techniques have different strength and weakness. The implication is that no single method is always best for all situations. Given an awareness of this dilemma, this research has adopted the use of triangulation strategy (e.g. Jick, 1979). Triangulation argues for the need to appropriately combine different methodologies to study a given phenomenon (Denzin, 1978). The concept of triangulation is based on the assumption that any bias inherent in particular data sources and research methods would be reduced or neutralised when used in conjunction with other data sources and research methods (Jick, 1979). Through triangulation, different methods are used to corroborate the same facts thus improving accuracy and providing the researcher with more confidence of the results (Das, 1983).

Before presenting the triangulation strategy adopted to ensure the validity of this research within the context of the nested approach used in this research, key terms will be described.

4.6.1 Validity

'Validity' is concerned with the extent to which the research findings present a true picture of what is being studied and what is really happening in the situation (Cunningham, 1988; Hussey and Hussey, 1997). Among the different types of validity, those most often used are construct validity, external validity and internal validity (e.g. Yin, 1994). Construct validity refers to 'the establishment of correct operational measures for the concepts being studied' (Yin, 1994, p. 33); external validity refers to 'the possibility of extrapolating the results obtained from a sample to other element, under different conditions of time and place' (Royer and Zarlowski 2001, p. 147) and internal validity 'consists in ensuring the relevance and internal coherence of the results in line with the researcher's stated objectives' (Royer and Zarlowski, 2001, pp. 147–148).

A single case study approach was used to conduct this research (see Section 4.3.2). Two criteria – validity and reliability – are most often used in evaluating the quality of the case study research (e.g. Yin, 1994). The important emphasis here is that the quality evaluation of this research is that the researcher takes precautions to improve validity and reliability, rather than testing and assessing the research's validity and reliability (Allard-Poesi *et al.*, 2001). Reliability is to be considered in the next subsection.

To ensure construct validity, this research triangulated the data collection process as much as possible. The data collection included a research focus phase and a case study phase which contained exploratory and action research elements (see Section 4.2). In the research focus phase, a number of general management and construction-specific literatures were reviewed and synthesised (see Section 4.4.1). In the longitudinal 22-month case study phase, the data were collected by carrying out interviews, reviewing company documentation, presenting and debating the findings at a workshop and carrying out an action research intervention.

Internal validity was strengthened by offering integrated research questions, hypotheses, a concept model and gap analysis framework which provides internal focus and cohesion to the results.

To ensure external validity, an explicit research design was developed for a single case study, including an articulated sampling strategy for the case study selection (sample size, classification of organisations) and sampling strategy for interviews (see Section 4.3). This explicit research design allows other researchers to understand how the results were produced, and to challenge, or confirm, the results by being able to replicate the research process in other case studies.

4.6.2 Reliability

Reliability is information on whether the instrument is collecting data in a consistent and accurate way. Simon and Burstein (1985) state that 'reliability is essentially repeatability – a measurement procedure is highly reliable, if it comes up with the same result in the same circumstances time after time,

Table 4.4   The tests for validation of this research.

Tests	How it is achieved

Validity
  Construct validity
  (Data collection)
	Data collection triangulation
	  Data were collected through multiple means, including a research focus phase, and a case study phase contained an exploratory phase and an action research phase (see Section 4.2).
	  In the research focus phase, data were collected through a number of general management and construction specific literatures (see Section 4.4.1).
	  In the case study phase, data were collected through multiple sources, including interviews, company documentation, company workshop and interventions (see Sections 4.4.2, 4.4.3 and 4.4.4).

External validity	Research design An explicit research design allowed other researchers to understand how to use it in other case studies (see Section 4.3).

Internal validity
  (data analysis)
	Research design
	  Integrated research questions, hypotheses, a concept model and gap analysis framework, provided internal focus and cohesion to the results.
	A longitudinal case study
	  A longitudinal 22-month case study offered a rich picture which reduced the risks of misjudgement of the truth value of data (see Section 4.3.4).
	Research design
	  An explicit research design which other researchers can follow (see Section 4.3).

Reliability	Case study protocol
	  The use of the semi-structured interview protocol by asking the same questions to five respondents enhanced reliability of the exploratory phase of the research (see Section 4.4.2).
	Action research process
	  An explicit action research methodology which other researchers can follow (see Section 4.5.4).

Representativeness	Sampling strategy The use of sampling strategy for the sampling design (sample size and classification of the firms) to select a suitable case study company and interviewees enhanced representativeness of the data (see Sections 4.3.2 and 4.3.3).
Generalisability	Case study design The sampling strategy enabled a representative small construction professional practice to be selected (see Section 4.3.2).

even employed by different people'. This definition has been extended by Yin (1994, p. 36) who states that reliability is the extent to which a test or procedure produces similar results under constant conditions on all occasions.

The reliability of this research was strengthened in three ways. First, the overall research design has been explicitly articulated and, therefore, can be replicated by future researchers. Second, in the exploratory phase, a semi-structured interview protocol was used. The questions within this protocol were based on the research hypotheses (see Section 4.4.2). The same protocol was used for all five interviewees. The action research phase was unique to the case study company and concentrated on a specific intervention. This part of the research, therefore, is not repeatable. Finally, the methodology explored in the data analysis has been described to a design where other researchers can both trace this researcher's analysis of the primary data and undertake their own analysis of the same data.

4.6.3   Representativeness

In a very broad sense, representation means 'the structure composed of the beliefs, values and opinions concerning a specific object, and the interconnections between them' (Allard-Poesi *et al.*, 2001, p. 351). To ensure representativeness, the researcher paid attention to robust the single case study design by designing a careful sampling strategy when selecting the case study firm (sample size, classification of organisations) (see Section 4.3.2) and by designing an appropriate sampling strategy for the interviews (see Section 4.3.3).

4.6.4   Generalisability

Generalisability has been defined as 'the extent to which you can come to conclusions about one thing (often a population) based on information about another (often a sample)' (Vogt, 1993, p. 99). The weakness of the case study approach is that the results cannot be generalised beyond the case study firm. This research adopts the position set out by Yin (2003, p. 39) in that the results are generalised to theory (which is analogous to the way in which scientists generalise from experiments to theory) rather than to the wider population of small construction professional practices.

The above discussions are summarised in Table 4.4.

4.7   Summary and Link

This chapter has set out the methodology used in this research. The next chapter presents the key results of the exploratory phase of the case study.

5

Case Study – Exploratory Phase

The aim of this chapter is to present and critically discuss the key findings from the exploratory phase of the case study (see Section 4.3.4). The concept model is used as an analytical framework to identify and distinguish the key variables for 'successful' and 'unsuccessful' innovation (see Section 3.2). To enable this, the chapter first develops a case-study-specific 'vocabulary' of concepts, namely knowledge, innovation, relationship capital (RC), structure capital (SC) and human capital (HC). Second, using this vocabulary, seven innovations which have taken place in the case study firm are analysed. The chapter is organised as follows:

(1) the background of the case study company is described (Section 5.2);
(2) the ArchSME perception of knowledge, innovation, RC, SC and HC, as described by the interviewees, is set out (Section 5.3);
(3) the company innovations identified by the interviewees are introduced (Section 5.4);
(4) the innovations categorised as being explorative in nature are described and analysed (Section 5.5); and,
(5) the innovations classified as being exploitative in nature are discussed and evaluated (Section 5.6).

5.2   Background of the Case Study Company

ArchSME is an architectural practice located in south Manchester in the northwest region of England. The practice was founded by the chairperson in 1991. ArchSME's principal markets are the Manchester city central and suburban residential sectors: varying from one-off commissions from regional clients to repeat business from national house builders. ArchSME currently has three principal clients. Two clients are large organisations (more than 251 staff), whilst one is a micro-organisation (less than 10 staff) (see Section 4.4.3). The clients all come from the private sector. Senior management believe the reasons that these clients remain with ArchSME is that it has the ability to deliver a good quality service, talented teams and built productive, ongoing client relationships.

A key external pressure for ArchSME (as it perceives) is that its national clients are demanding that it is accredited with ISO 9001 and/or Investors in People (IiP). ArchSME recognises that this demand for accredited status provides opportunities to access the public sector market, whilst ensuring that they remain the leaders in its current target markets. The ArchSME management believes that IiP would practically benefit the organisation by providing a framework to incorporate better business practice, and develop and maintain a 'winning' team (Lamb, 2003). In February 2003, ArchSME was granted an IiP accreditation. ArchSME was working during the case study period to achieve ISO 9001 accreditation.

In May 2002, ArchSME relocated from their long-standing rented accommodation in Hale, and purchased its own office block in Altrincham. The new office is approximately 5 miles from its previous office. The reason for the relocation was that it supported the first step in its strategy to grow the size of the practice. The new building has extra space (currently rented out to another firm) to 'expand into' if needed at a later stage. The move gave the company an opportunity to advertise its growth and to communicate to the marketplace its seriousness in becoming a very successful architectural practice with the capability and capacity to compete with larger regional and national practices.

Over the past 5 years, the practice has grown significantly with an increase in turnover from £0.3 million in 1999 to £1.6 million in 2003 (see Figure 5.1). Employee numbers have grown: 12 in 1999, 34 in 2002 and 40 in 2003. Turnover per employee increased from £25 000 per employee in 1999 to £40 000 per employee in 2003. Pre-tax profit levels have remained comparatively low compared to the growth in turnover as a result of an explicit policy to invest in company growth (for example, the purchase of the new office in 2002).

The practice is a limited company, and is owned and managed by a team of three equity directors – a chairperson, a managing director and a

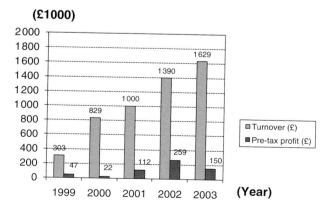

Figure 5.1   ArchSME's turnover and pre-tax profit in the last five financial years.

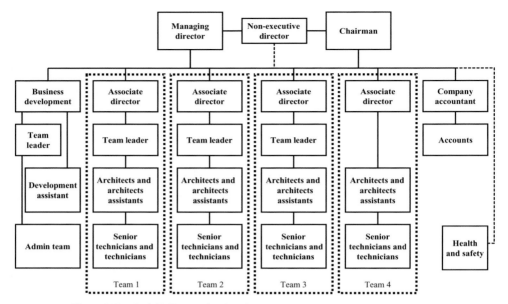

Figure 5.2   ArchSME organisational structure.

non-executive director. The organisation and management structure of the company is shown in Figure 5.2. There are four teams and two support units within the practice. The support units are based on functional expertise: financial accounts and business development. The four teams are organised as individual profit centres responsible for its own marketing, professional service development and delivery. Each team is made up of an associate director, a team leader (except team 4) and a number of architects and technicians. Team 4 undertakes minor works only. Only the associate directors report to the managing director.

The way that work is brought into the firm is shown in Figure 5.3. Work comes from two principal sources: clients and contractors. The potential commission is managed by an associate director initially, before reporting it to the senior management board, which comprises the directors and associate directors. The acceptance of the commission is made by the managing director in the management meeting. An appointed team manager (an associate director) goes back to his or her team and assigns project team members to

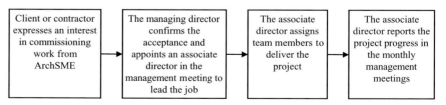

Figure 5.3   The commissioning and delivery of work process in ArchSME.

deliver the project. Progress on the project is reported at subsequent senior management board meetings.

The workflow with the company is described as follows, using team 1 as an example. There are six staff in the team: an associate director, a team leader, an architect (the job runner), an architectural assistant, a senior technician and a technician. The associate director is the project team manager and assigns tasks to team members. The associate director and the team leader are responsible for the delivery of the service to clients. The architect (the job runner) establishes detailed client and regulatory requirements for the job. The architectural assistant and two technicians are responsible for the preparation of drawings and related technical documentation as instructed by the team leader and architect and as required by British Standards, building regulations and the ArchSME-specific CAD standards. All the team 1 members are located in the same block in the office. The teamwork often takes place in an informal way, such as 'corridor' discussions and ad hoc meetings.

5.3   Case Study Firm Perception of Knowledge, Innovation, HC, SC and RC

5.3.1   Definition of knowledge

The variables making up ArchSME's perception of knowledge is set out in the cognitive map shown in Figure 5.4. The following discussion is supported by references to the cognitive map (for example, '8 3' refers to supplier level). This notation is used throughout this chapter.

The respondents viewed knowledge in a variety of ways depending on the level of resolution, be it at an individual level, company level or supplier level. At an individual level, knowledge was conceived as the 'ability' (8 1 1) to perform a task competently. Respondent B, for example, stressed that knowledge was:

> the ability to carry out your job.

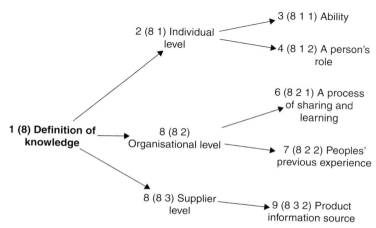

Figure 5.4   Knowledge cognitive map.

Knowledge was also seen as the knowledge of 'a person's role' (8 1 2) and how that role interacts with other roles within the firm. Respondent A, for instance, emphasised that knowledge:

> is knowing your role...[and]...knowing your place in the team.

At an organisational level, it was found that knowledge is embedded within and across people. It was evident in 'peoples' previous experience' (8 2 2) variable. Individual knowledge is seen as the building blocks for sharing and learning within the organisational community. 'A process of sharing and learning' (8 2 1) was emphasised by Respondent D, who expressed that knowledge is:

> the key, we cannot develop, unless we introduce knowledge and share knowledge within the rest of my team. It's actually the key to what we do – sharing.

This tacit view of organisational knowledge was supported by Respondent E, who described knowledge as:

> what you've learnt personally or tacitly from someone else...passed on knowledge.

The development and sharing of knowledge is seen as specific to the firm and a potential source of unique, added value. Respondent D argued that:

> it's very difficult to put what we do, or describe what we do to other people within the industry. Our knowledge is developed in-house, and then we share the product.

The conceptualisation of knowledge as being tacit at an individual and organisational level migrates to a more explicit, 'product' view of knowledge at a supplier level. The supplier was as a 'product information source' (8 3 1), with Respondent B, for instance, stressing that:

> the supplier is able to give you information [on a specific product] you need to put on the task at the time.

In summary, a 'process' view of knowledge is prevalent within ArchSME activity, tacit understanding and sharing of knowledge and roles specific to individuals and firms. Knowledge is not seen as an 'asset' which is encoded and stored in databases. Knowledge is a living, personalised phenomenon – not 'blocks' of data and information.

5.3.2   Definition of innovation

The variables making up ArchSME view of the definition of innovation is set out in the cognitive map shown in Figure 5.5.

The respondents viewed innovation in different ways depending on the level of resolution, be it at an individual level, company level or supplier level.

At an individual level, innovation is seen as 'a new idea' (9 1 1). Respondent E, for example, argued that innovation is a:

> new product or a new way of doing things.

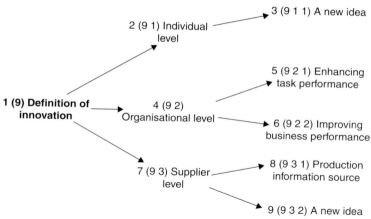

Figure 5.5   Innovation cognitive map.

This concept of newness was extended to encompass individual creativity. Respondent A, for instance, stated that innovation is:

being able to think unlike your colleagues or unlike people before you.

At an organisational level, innovation is seen as 'enhancing task performance' (9 2 1). Respondent B, for example, emphasised that innovation is:

using the product that is better suited to performing the task.

This perception was extended to explain that innovation at an organisational level needed to 'improve overall business performance' (9 2 2). Respondent E, for instance, argued that innovation is:

a new way of doing things to improve the business . . . for development.

At the supplier level, innovation was conceived as being the same as an 'individual' innovation in terms of a new idea which has the benefit of input from relevant people in the supply chain. This was evident in 'a new idea' (9 3 2) variable and was demonstrated by Respondent D, who described innovation as:

a one good idea. We then may need to develop that. We then may need other people's knowledge, other people's input from the industry.

It was found that the supplier as 'product information source' (9 3 1). Respondent B stated that innovation is:

looking for the supplier chain, all of the suppliers, to give you information to make sure that it is an innovative product, and add something new will be carried out on your job. That's new compared to the previous things you give them.

In summary, innovation is seen to apply 'a new idea' to enhance the task and overall performance within ArchSME. The source of ideas is more likely to be from personal creativity or the outcome of social interaction, rather than learned from codified sources such as trade journals or books.

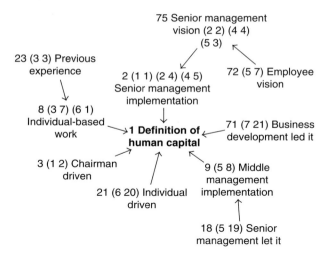

Figure 5.6 Human capital cognitive map.

5.3.3   Definition of human capital

The multidimensional nature of HC is portrayed in the cognitive map shown in Figure 5.6.

The respondents viewed HC as being synonymous with the staff of ArchSME. Respondent A, for example, commented that:

the company is only as good as its people.

Individual ability to create and implement ideas depends heavily on their ability to mobilise and synthesise appropriate bodies of expertise and experience to a specific application domain. The 'previous experience' (3 3) was seen in the 'individual-based work' (3 7; 6 1) variable. The ability of staff to create ideas was evident in the 'senior management vision' (2 2; 4 4; 5 3) and 'employee vision' (5 7) variables. The ability of staff to implement the ideas was demonstrated in the 'senior management implementation' (1 1; 2 4; 4 5), 'middle management implementation' (5 8), 'senior management led it' (5 19) and 'business development led it' (7 21) variables. The combination of the idea creation and implementation was clear in the 'individual driven' (6 20) and 'chairman driven' (1 2) variables.

People are seen as the sources of information. Respondent D, for example, asserted that:

the information source is the people . . . rather than our product; not documents.

The way information is collected is taken to be through people interaction. Respondent D, for instance, emphasised that:

It's by just talking to people . . . that's how information is collected in the practice.

Social interaction of this nature is an important mechanism for knowledge sharing. Respondent C, for example, stressed that:

During sharing knowledge with my colleague, so I got this idea that we have this new material.

The perception was extended to explain that a process view of knowledge within the staff is seen as specific to the firm. Respondent D, for example, emphasised that:

our industry, what we do, isn't the sort of things you can put down on the database, because what we do, everything we design, should be new, should be an idea to present, to develop.

In summary, HC within ArchSME is seen as being very much synonymous with the knowledge and skills of individuals, and the ability of individuals and teams to mobilise and synthesise this knowledge and skills to specific application domains.

5.3.4   Definition of structure capital

The variables making up SC is set out in the cognitive map shown in Figure 5.7.

The SC within ArchSME was principally viewed as being the formalised organisational structure and document repositories which encourage and support people to share their knowledge. The process view of knowledge was captured in the recent company restructuring, including the 'management meeting' (1 26; 5 11), 'quarterly office meeting' (1 23; 4 6) and 'annual staff appraisal' (2 10; 3 9) variables. This was confirmed by Respondent D, who expressed that:

by looking at pictures, ideas and sharing and that was done informally. But we still need structures to be in place to ensure we are sharing that information.

Respondent B, for instance, described that:

you get meetings every so often to present information and to share where the company standard is at any given time.

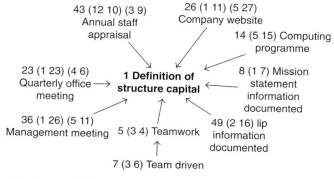

Figure 5.7   Structure capital cognitive map.

The SC was also seen as the team structure to perform the job, from idea creation to delivering the service. This was evident in the 'team driven' (3 6) and 'teamwork' (3 4) variables.

The asset view of knowledge is apparent within the SC. This was illustrated in the 'mission statement information documented' (1 7), 'IiP information documented' (2 16), 'computing programme' (5 15) and 'company website' (1 11; 5 27) variables. Respondent C, for example, emphasised that:

> The information sources need to be accessible. Now we have a company manual and the structure within the company is all in there.

In summary, SC is seen as the organisational context in which a process view of knowledge creation by staff can take place, and knowledge content, from an asset perspective, encoded within accessible documentation.

5.3.5   Definition of relationship capital

The key variables making up RC are presented in the cognitive map shown in Figure 5.8.

The RC is seen as creating and maintaining good relationships with colleagues, suppliers and company external business advisers. The importance of 'good relationships' was demonstrated in the 'good relationships with clients' (6 4), 'good personal relationships with suppliers' (6 8) and 'good relationships with colleagues and suppliers' (1 10) variables. Respondent C, for example, described the inherent momentum of relationship building by saying:

> When you're dealing with clients, you develop a relationship.

The RC is also seen as a key source of information. Through people interaction, the information is collected. This was seen in the 'informal team

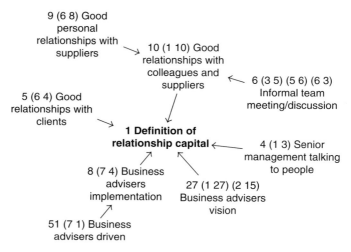

Figure 5.8   Relationship capital cognitive map.

meeting/discussion' (3 5; 5 6; 6 3) and 'senior management talking to people' (1 3) variables. Respondent D, for instance, described how senior management collected the information in the architectural practice:

> Architecture is a very small world. Although a lot of companies are competitors and/or consultants . . . you still talk to people a lot. We meet some friends from different organisations, especially the senior management here have a lot of contacts with other architects and understanding how they view us, it's by just talking to people . . . that's how information is collected in the practice.

It was found that business advisers have an important influence on idea creation and implementation within the firm. This was evident in the 'business advisers vision' (1 27; 2 15), 'business advisers driven' (7 1) and 'business advisers implementation' (7 4) variables. The business adviser implementation was captured by Respondent E, who stated that:

> [Business advisers] went to the open day and said what kind of courses have you got and they came away and asked what kind of courses they wanted and enrolled.

In summary, RC is seen as the creation and maintenance of enduring internal and external relationships. These relationships are both a rich source of ideas and the arena for appropriate innovation to ensure successful problem solving.

5.4   Description of Identified Company Innovations

Seven innovations were identified by the respondents as being significant firm-generated innovations over the last 2 years: four being deemed successful and three unsuccessful. Each innovation is briefly described below.

The development of the ArchSME mission statement (innovation 1), the securing of IiP accreditation (innovation 2), the flow of new novel designs (innovation 3) and the company restructure (innovation 4) were identified as being significant firm-generated innovations over the last 2 years which were successful.

> *Innovation 1: mission statement* is a statement that captures an organisation's purpose, customer orientation and business philosophy. ArchSME's mission statement is 'to be recognised as the leading north west [England] design house dedicated to achieving working relationships which result in excellent architectural solutions'. This mission statement was created and introduced to the company in October 2002.
> *Innovation 2: IiP* is the UK national standard which sets out a level of good practice for training and development of people to achieve business goals (www.investorsinpeople.co.uk). ArchSME secured accreditation in February 2003, after a 1-year period of preparation.
> *Innovation 3: new designs* are novel forms of layout and structure. ArchSME have, in its view, consistently produced innovative designs for new buildings.

Innovation 4: company restructure is the way in which ArchSME is organised and work coordinated to ensure successful delivery of service to the client. The company was restructured in 2002 to meet general business needs and to prepare itself for IiP accreditation.

Respondents identified the introduction and subsequent failure of in-house seminars (innovation 5), the introduction of the new materials (innovation 6) and the Learndirect project (innovation 7) as being significant innovations over the last 2 years which failed.

Innovation 5: seminar is a type of meeting where ideas are exchanged on a specific topic. The identified seminars within ArchSME included IT, project briefing and marketing. Two to three representatives from each team are chosen by the associate director and sent to attend IT and marketing seminars. In the project-briefing seminar, a team is appointed by the managing director to present one of their projects to the other three teams. The seminars started in August 2002, and petered out by February 2003.

Innovation 6: new materials are the building components, materials or products that the company incorporates into its building designs for the first time.

Innovation 7: Learndirect project is funded by the UK government (www.learndirect.co.uk). This project aims to help people to develop their IT capability in getting easy access to information about what is available. Business advisers from the Learndirect project had an informal discussion with each member of ArchSME staff during an open day in September 2002. From this, each employee had a personal development plan (PDP) established. These PDPs have not been progressed or embedded within the ArchSME's appraisal system.

The research key findings indicate two types of innovation within the company: explorative innovation (see Section 5.5) and exploitative innovation (see Section 5.6). It is argued that firms achieve short-term success with explorative innovation (see Table 5.1, mode 1 and Figure 5.9, mode 1) and long-term success with exploitative innovation (see Table 5.1, mode 2 and Figure 5.9, mode 2). The classification of explorative and exploitative innovation is used to structure the following sections, and is justified below.

(1) *Explorative innovation* (mode 1) is viewed as innovation which focuses on client facing, project-specific problem solving. Explorative innovation activity heavily relies on the capacity, ability and motivation of

Table 5.1 Classification of explorative and exploitative innovation.

Types of innovation	Mode 1: Explorative innovation	Mode 2: Exploitative innovation
Successful innovation	Innovation 3: New designs	Innovation 1: Mission statement Innovation 2: Investors in People (IiP) Innovation 4: Company restructure
Unsuccessful innovation	Innovation 6: New materials	Innovation 5: Seminars Innovation 7: Learndirect project

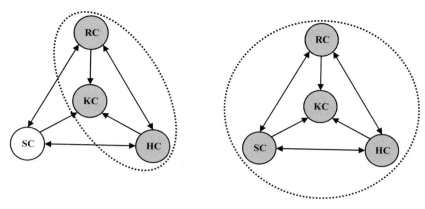

Mode 1: Explorative innovation   Mode 2: Exploitative innovation

Figure 5.9   Types of innovation.

ArchSME staff at an operational level to solve client problems and, in doing so, generates short-term competitive advantage (i.e. project specific). The outcome of this innovation focuses on effective and efficient delivery of services to satisfy current external project needs, but are often not embedded in the organisational SC due to the required management attention and company resources to make this happen being constantly focused on other current or future project-specific considerations. Explorative innovation activity is discussed in Section 5.5.

(2) *Exploitative innovation* (mode 2) is viewed as innovation which focuses predominantly on internal organisation and general client development activity which is not project-specific fee-earning activity. Exploitative innovation activity heavily relies on the capacity, ability and motivation of ArchSME senior management at a social level to improve organisational effectiveness and efficiency to generate sustainable competitive advantage. The distinctive feature of exploitative innovation (compared to explorative innovation) is that new phenomena, systems or structures are securely embedded in the SC of the firm. Exploitative innovation activity is discussed in Section 5.6.

The key proposition being made in this section is that the concept of exploitative and explorative innovation is an appropriate way of understanding knowledge-based innovation. The next section presents an analysis of explorative innovations.

5.5   Mode 1: Explorative Innovation Analysis

Two exploitative innovations were identified as being significant firm-generated innovations over the last 2 years. The successful explorative innovation was considered as new designs (innovation 3), whilst the unsuccessful

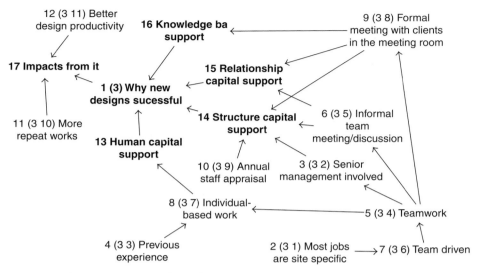

Figure 5.10   Successful explorative innovation (new designs) cognitive map.

one was the use of new materials (innovation 6) (see Section 5.4 for the description of innovation 3 and innovation 6). Both explorative innovations were identified by Respondent C; therefore, primary data are from this respondent only.

The key factors and interrelationships for the successful explorative innovation are given in Figure 5.10, and for the unsuccessful explorative innovation are shown in Figure 5.11. These form the basis, along with appropriate extracts from the interview transcripts, for the following discussion.

5.5.1   Human capital

The HC was found to be embedded within the capacity, ability and motivation of staff. Individual ability to compete successfully depends heavily on their ability to mobilise and synthesise bodies of expertise and experience in order to create knowledge that satisfies client demands. This was evident in the 'individual-based work' (3 7; 6 1) variable. In successful explorative innovation, the 'previous experience' (3 3) was seen as being important for knowledge workers in performing their works. This was captured by Respondent C, who stated that:

> Design work is like showing clients what we've done before, showing clients other schemes, showing clients how it works previously. It's like showing clients the different designs we can do. (Innovation 3: new designs)

In unsuccessful explorative innovation, the adopted idea (a new material) had to be used before it was shown in the 'the recommended product been used before' (6 5) variable. The previous experience was seen to give the staff

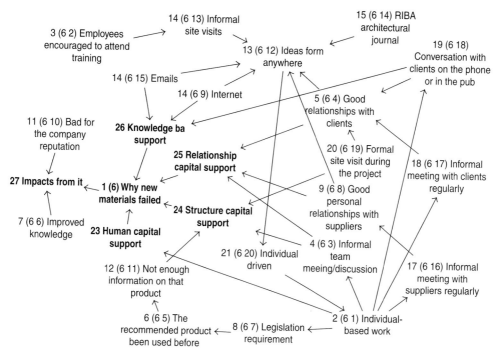

Figure 5.11   Unsuccessful explorative innovation (new materials) cognitive map.

and the client confidence in adopting new idea. This was demonstrated by Respondent C, who expressed that:

> I have never [been the first one to use a new material], but it must be difficult to use that new material if it has never used before, to be able to have confidence in it. (Innovation 6: new materials)

The 'most jobs are site specific' (3 1) reality encouraged staff to be 'self-motivated' in that they are directly responsible for the creation and use of an idea within a project-specific situation. This was described by Respondent C, who stressed that:

> Most jobs are site specific any way. So ideas need to change, evolve for specific clients, for specific sites . . . . (Innovation 3: new designs)

The key distinction between successful and unsuccessful explorative innovations, from an HC perspective, was the 'social' or 'operational' nature of the knowledge being applied to a specific innovation. 'Operational' activity is where the focus is on solving project-specific problems. These projects are either 'external', fee-earning projects or 'internal' but specific client-driven projects. 'Social' activity is where the focus is on generating non-project-specific innovation which build up general organisational capability and deeper client relationship over the medium to long term.

In successful explorative innovation, the application domain was a specific project, where knowledge gleaned from 'social' or 'operational' levels (see Section 5.5.3) was appropriately filtered and configured to meet the unique needs of the project. The creation and application of knowledge at an operational level was clear in the 'team driven' (3 6) variable and was identified by Respondent C, who stated that:

> Initially ideas are always from within the team, and then we focus on integration with other teams within the office. (Innovation 3: new designs)

In the cases of unsuccessful explorative innovation, the creation of ideas from individual creativity was seen in the 'individual driven' (6 20) variable and was captured by Respondent C, who stressed:

> ... [using new materials] are down more to an individual basis ... ideas ... might come from individual, from me; might come from a supplier or might come from a client's suggestion. (Innovation 6: new materials)

It was found that the 'ideas from anywhere' (6 12) variable was particularly pertinent in unsuccessful explorative innovation. Ideas might come from the 'internet' (6 9), 'emails' (6 15), 'good relationships with clients' (6 4), 'good personal relationships with suppliers' (6 8), 'Royal Institute of British Architects architectural journal' (6 14) and 'informal site visits' (6 13). Knowledge workers learn from such external or internal sources and generate 'background' knowledge, but this knowledge does not directly and immediately feed into current projects. Respondent C, for example, articulated that:

> Recently we have been looking at a large high rise apartment scheme, visits around Manchester, and looking at apartment schemes to look at what other people are doing to formulate some ideas for what we should be doing in the future. (Innovation 6: new materials)

In unsuccessful explorative innovation, ideas were socially derived but were not project specific at the time of their inception (see Section 5.5.3).

In summary, HC for explorative innovation was found to be embedded within the capacity, ability and motivation of staff. The key distinction between successful and unsuccessful explorative innovations, from an HC perspective, was the 'social' or 'operational' nature of the knowledge being applied to a specific innovation. In successful explorative innovation, the application domain was a specific project, where knowledge gathered from whatever source ('social' or 'operational' levels) was appropriately filtered and configured to meet the unique needs of the project through the team structure. In contrast, unsuccessful explorative innovation was characterised by socially derived knowledge which was not adequately transformed to meet the need of a specific project, and was thus incompatible with the operational pool of knowledge being used.

5.5.2   Structure capital

The principal locus of SC was found to be the team structure and associate team working.

The SC within 'teamwork' (3 4) was seen as being important in progressing specific project issues. At an operational level, the 'teamwork' (3 4) was captured in activities including the 'formal meeting with clients in the meeting room' (3 8), 'formal site visit during the project' (6 19), 'informal team meeting/discussion' (3 5; 6 3) and 'team driven' (3,6) variables. The way of the teamwork was described by Respondent C, who stressed that:

> So for a specific product [the team arranges] to look at that the product. The team working with that product will go and see that product. (Innovation 6: new materials)

The role of senior management in doing work through the team structure at an operational level was articulated in the 'senior management involved' (3 2) variable. It was evidenced by Respondent C, who stated the importance of senior management in the teamwork:

> Senior management will sometimes be part of these meetings. Sometimes they go down to discussing individual jobs, and whether or not [clients] want to get a senior manager involved. (Innovation 3: new designs)

In contrast, unsuccessful explorative innovation was found to have its foundations in individually created ideas derived from his or her 'social' RC which were inappropriate for the specific project needs, and which were pursued relatively independently of the team. The role of the individual in doing work at an operational level was articulated in the 'individual-based work' (6 1) variable and was captured by Respondent C, who stressed the early devolvement of responsibility to junior staff:

> A lot of younger, less experienced members of staff, get a quite lot of responsibility. [Innovation activity] doesn't necessarily always need senior management. (Innovation 6: new materials)

Although 'the recommended product been used before' (6 5) or the product had met 'legislation requirement' (6 7), 'not enough information on that product' (6 11) was identified as the key obstacle in unsuccessful explorative innovation. Respondent C, for example, asserted that:

> It's generally a sales problem … because it didn't provide enough information about [its firm's] products. (Innovation 6: new materials)

In unsuccessful explorative innovation, the socially derived ideas did not have sufficient demonstrable benefit or momentum to become embedded in SC. Explorative innovation success or failure was found to be determined by the 'annual staff appraisal' (3 9) and 'formal site visit during the project' (6 19) activities. The lack of 'quantitative' innovation performance measurement system was captured by Respondent C, who commented that:

> There isn't really a structural reward system [for rewarding successful innovation] in place for us as I am aware of, but I think like Christmas bonus etc. If we're doing well, performing well, we get feedback in that way. There is [the annual staff appraisal]. (Innovation 3: new designs)

In summary, the principal locus of SC was found to be the team structure and the group dynamics within these teams. Successful explorative innovation was found to have enduring senior management support from inception through to implementation, and supported by an enabling team structure which stimulated and developed team-based ideas at an operational level. In contrast, unsuccessful explorative innovation was found to have its foundations in individually created ideas derived from his or her 'social' RC (see Section 5.5.3) which were inappropriate for the specific project needs, and which were pursued relatively independently of the team. These ideas did not become embedded at an operational SC level. In successful and unsuccessful explorative innovation, there was found to be a lack of a 'quantitative' innovation performance measurement system to determine the success of innovation activity. Intuition and collective perceptions determine success or failure of an innovation. The limitation of relevant and up-to-date information within the structure is seen to be a further, key obstacle to explorative innovation success.

5.5.3   Relationship capital

The RC was evident within ArchSME and was characterised as being at internal, client and supplier interaction domains of activity.

The RC within 'an internal' context is seen as being important in nurturing communication and cohesion across vertical and hierarchical levels. This was shown in the role of 'informal team meeting/discussion' (3 5; 6 3) which was described by Respondent C, who stated that:

> [Relationship capital is] quite dominant in our firm really. That's working in the team and teams change within the company. So we need to have close relationships between our colleagues within the practice, and also senior management and lower levels of staff to encourage, and things like that, to seek advice when we need it. (Innovation 3: new designs)

At a client interaction level, RC is viewed as being important in terms of 'operational' interaction to progress specific project issues and 'social' interaction to forge and replenish non-project-specific relationships with clients. 'Formal meeting with client in the meeting room' (3 8) and 'formal site visit during the project' (6 19) were identified as being key operational RC mechanisms and was illustrated by Respondent C, who explained that:

> [The activities carried out to support the new designs] were formal presentations and meetings with the clients. (Innovation 3: new designs)

The social interaction aspects of knowledge workers and clients interaction were captured in activity including 'informal meeting with clients regularly' (6 17) and 'conversation with clients on the phone or in the pub' (6 18). Respondent C, for example, articulated that:

> I go off and meet clients on a regularly basis. Then we just cover a lot of things specifically . . . generally to just talk about things. (Innovation 6: new materials)

It was found that having good relationships with clients have significant influence in the application and acceptance of new ideas. Respondent C, for example, articulated that:

> I don't think I can remember specific cases where we have lost clients... because we have such good relationships with clients anyway. We are quite highly judged by the clients. We did quite a lot to make sure we look after the clients. So probably it is more a level of tolerance with us than with other companies. We can potentially make a few more errors to potentially make improvements afterwards. (Innovation 6: new materials)

The good relationship with clients also had an input into the company marketing. This was stressed in the 'more repeat works' (3 10) variable and was captured by Respondent C, who articulated that:

> We don't advertise very much. It's mainly repeat work we get anyway. So we don't need to compete really. (Innovation 3: new designs)

Interaction between knowledge workers and suppliers was emphasised in the 'good personal relationships with suppliers' (6 8) variable. Again, the distinction between 'operational level' and 'social level' interaction was evident. At an 'operational' level, Respondent C described the benefits in good relationship with suppliers:

> I have very good relationship with at least five suppliers. If I want it to show the client a new product...I will get the supplier to provide a sample which is specific to the design we are talking about. (Innovation 6: new materials)

In contrast, at a 'social' level, the 'informal meeting with suppliers regularly' (6 16) variable was evidenced by Respondent C, who described that:

> Me, having informal meeting with much suppliers every few weeks if they have new products to show and, ordinarily, the supplier will want to come in and talk it through. Certainly the company wants to do that. (Innovation 6: new materials)

It was found that the good supplier operational RC was instrumental in generating the enabling conditions for creative action. This position was captured by Respondent C, who described that:

> After developing the relationship with the supplier, you can ask them for [new material] information. You can find out more information if those suppliers are trusted. (Innovation 6: new materials)

The logic of pursuing both 'operational' and 'social' RC was that social RC developed the supportive context within which operational relationships could prosper. This aspiration was commented on by Respondent C, who argued that:

> If you have a good social relationship with clients, with consultants, it means you have good working relationship with them as well. (Innovation 3: new designs)

The social RC exposes knowledge workers to new possibilities to feed into operational RC at a project-specific level at a future date. Respondent C, for example, articulated that:

> We can learn more about how the detail can be done correctly next time etc. (Innovation 6: new materials)

In summary, RC is seen as the results of internal, client and supplier interactions. Two broad types of RC were grouped. First, 'operational RC' was to progress specific project needs. Second, 'social RC' was to forge and replenish non-project-specific relationship with others at work. It was found that social RC has a significant effect on feeding operational relationship at a specific project level at a future date.

The successful explorative innovation was found to have 'operational' and 'social' RC sources which were fed into project-specific innovation needs. In contrast, unsuccessful explorative innovation was underpinned solely by 'social' RC sources which did not meet project-specific innovation needs.

5.5.4   Knowledge capital

The knowledge capital where HC, SC and RC were brought together within explorative innovation was distinguished as being located in 'social' and 'technical' contexts.

In a 'social' context, KC was seen to stimulate interaction and collective 'process-orientated' knowledge creation and conversion. In successful explorative innovation, the 'company environments' (such as office layout and meeting rooms) provided the social context to support team activity. It was evident in 'formal meetings with clients' (3 8) variable. Respondent C, for example, described the importance of the office layout in successful explorative innovation:

> All teams interact because of the office. The office is configured, so, for example, different resources and different floors and different people are configured. So everybody have to cross them in the office to see other people in their daily routine. So it is not about the people in the individual offices. They don't see other people during the day. (Innovation 3: new designs)

In unsuccessful explorative innovation, the public house and telephone conversations were found to be the basis for a social context to support individual activity and were evident in the 'conversation with clients on the phone or in the pub' (6 18) variable. Respondent C, for example, stated the way he interacted with people:

> Telephone conversations, conversations in the pub and that kind of thing. (Innovation 6: new materials)

In a 'technical' context, KC was seen to support the search for external knowledge and sharing of 'asset-orientated' knowledge. A 'technical' context view of KC within explorative innovation was seen to give an alternative to a 'social' context. Specifically, the importance of information technology (IT) such as 'the internet' searches (6 9) and 'emails' (6 15) was evident. The internet was identified as important technology for the information gathering and was captured by Respondent C, who noted that:

> A lot of people get their updates from the architecture journal from RIBA, providing suggestions, new product etc. There is normally a link to that website. (Innovation 6: new materials)

The use of email technology to share knowledge within the practice was evidenced by Respondent C, who stressed that:

> Quite often people who have been on seminars will provide a report, a formal type of report which is emailed to everybody. (Innovation 6: new materials)

However, there was no evidence that project-driven innovation was explicitly or adequately captured into the SC for subsequent retrieval and use in other projects by the same team or by teams.

In summary, KC is seen as the focal or integrating nexus in which innovation takes place. Two broad types of the nexus were distinguished. First, in a 'social' context, KC stimulated interaction and collective 'process-orientated' knowledge creation and conversion. This took the form of office environments, which supported team activity, such as meeting rooms and office layout. Second, in a 'technical' context, KC supported the search for external knowledge and sharing of 'asset-orientated' knowledge. This took the form of internet searches and emails, respectively.

In successful explorative innovation, KC was associated with a combination of 'social' and 'technical' contexts, particularly when KC was channelled to project-specific, operational activity. In contrast, unsuccessful explorative innovation was seen to be brought about when the KC was limited to a 'technical' dimension, as it tended to be located at an individual-driven social level (for example, 'surfing the net' for new construction technologies) and did not lend itself to team-based, socially constructed innovation activity.

5.5.5   Innovation outcomes

The outcome of successful explorative innovation resulted in effective and efficient delivery of services to satisfy current project-specific needs. This was evident in the 'better design productivity' (3 11), 'more repeat works' (3 10) and 'improved knowledge' (6 6) variables. Respondent C, for example, described how explorative innovation improved subsequent work productivity:

> Often when people have developed a successful detail, maybe a balcony that's worked really well, again it would get spread around the company. It improves productivity in future designs because you don't always want to redesign every part of building every time you do another building; it tends to try and make it more efficient for the design in the future. So we can almost use various parts of the building design again if it worked well in the first place. (Innovation 3: new designs)

Within this context, it was found that the outcome of explorative innovation was not embedded in the organisational SC, but located in individual SC.

The negative impact from unsuccessful explorative innovation was that it could damage ArchSME reputation, identified in the 'bad for company reputation' (6 10) variable. This was evidenced by Respondent C, who explained that:

> It's not good for the reputation but obviously if the product isn't working, especially we can work around it to see if we can change it and get back to the supplier to ask if we can change it. (Innovation 6: new materials)

In summary, the outcome of explorative innovation was found as focusing on effective and efficient delivery of services to satisfy current and/or future project-specific considerations/needs. It was found that the outcome of explorative innovation in terms 'best practice' was not captured and embedded in the organisational SC.

5.6   Mode 2: Exploitative Innovation Analysis

Five exploitative innovations were identified as being significant, firm-generated innovations over the last 2 years (see Section 5.4). The successful exploitative innovations were considered as the ArchSME's mission statement (innovation 1), the accreditation of IiP (innovation 2) and the company re-structure (innovation 4). Unsuccessful exploitative innovations were viewed as seminars (innovation 5) and the Learndirect project (innovation 7).

The key factors and interrelationships for successful exploitative innovation are shown in Figure 5.12, and for unsuccessful exploitative innovation are shown in Figure 5.13. These form the basis, along with appropriate quotes from the interview transcripts, for the following discussion.

5.6.1   Human capital

The HC for exploitative innovation was found to be principally embedded within the capacity, ability and motivation of senior management, and the level of employee participation in decision-making. Further, lack of time to implement ideas was found to be the critical obstacle for HC in supporting successful exploitative innovation.

The capacity, ability and motivation of senior management

The role of senior management involves the envisioning, creation and application of knowledge. The ability of senior management to generate new ideas was seen as a key aspect for exploitative innovation. The initial ideas for successful and unsuccessful exploitative innovation predominantly came from senior management and were evident in the 'senior management vision' (2 2; 4 4; 5 3) variable. The fact that the idea to restructure the company came from senior management was demonstrated by Respondent D, who said that:

> [The company structures] are actually structured . . . introduced and driven by senior management. They set the structure and then went down through the teams.

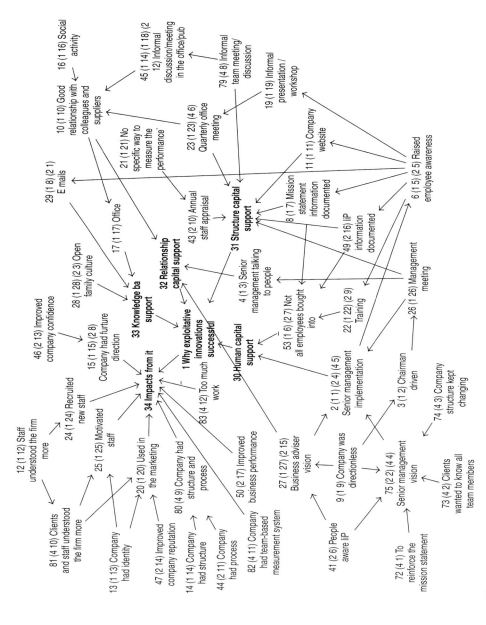

Figure 5.12 Successful exploitative innovation cognitive map.

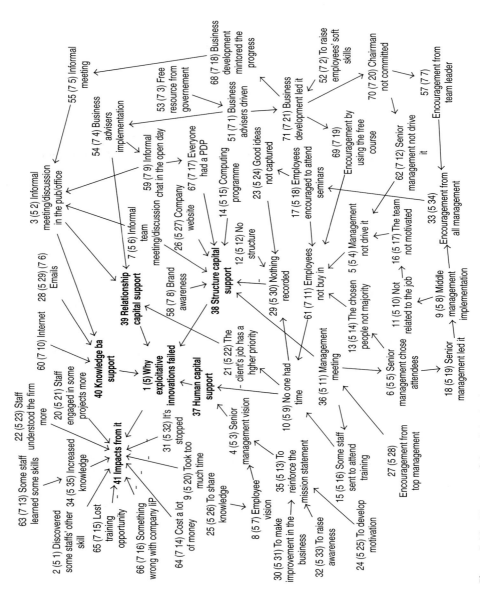

Figure 5.13 Unsuccessful exploitative innovation cognitive map.

It's always driven by senior management. It's not really a discussion point from there, from the other members. It's really a senior management issue ... director level. (Innovation 4: company restructure)

In unsuccessful exploitative innovation, the some of the idea to use seminars to share project information between teams coming from senior management was emphasised by Respondent E:

This initial idea came from [senior management] ... trying to increase our tacit knowledge throughout the company because we have a big problem with communication. So we tried to improve it then by using the project seminars. (Innovation 5: seminars)

In this context, the ability to scan and sense external and internal market stimuli and to make appropriate internal responses appeared to come from the senior management level. The awareness of the external market demands by the senior management was found to be reactive in nature. It was shown in the 'people aware IiP' (2 6) and 'clients wanted to know all team members' (4 2) variables. Respondent B, for example, stated that how senior management sensed the need for IiP:

It was a couple of years ago ... directors attended some business meetings in which it was stated that a lot of people are aware of the importance of getting Investors in People accreditation. (Innovation 2: Investors in People)

The idea for the success of exploitative innovation was found to meet ArchSME internal organisation needs and/or to develop general client relationship activity. The IiP (innovation 2), for example, was demonstrated in the 'company was directionless' (1 9), 'to reinforce the mission statement' (4 1; 5 13) and 'company structure kept changing' (4 3) variables. This was evidenced by Respondent A with respect to the strategic focus within rationale for the need for a mission statement to respond to a lack of ArchSME in the comment that:

The company is very much ... directionless ... we didn't know where we are going ... (Innovation 1: mission statement)

The ideas which stimulated subsequently unsuccessful exploitative innovation were found to have primarily been driven by individual needs. This was seen in the 'employee vision' (5 7), 'to share knowledge' (5 26), 'to develop motivation' (5 25), 'to raise awareness' (5 33), 'to make improvement in the business' (5 31) and 'to raise employees' soft skills' (7 2) variables. The Learndirect project (innovation 7), for example, was response to skills shortages, as emphasised by Respondent E:

It was a new idea to try to raise the skills. Instead of being professional qualified it was more about developing soft skills, like time management or managing meetings. So we wanted to develop their softer skills. (Innovation 7: Learndirect project)

The motivation of senior management to implement the innovation (see Section 5.6.2 for the description of senior management implementation) appeared important in determining whether or not exploitative innovation was successful. The need for dedicated top management was identified by the 'chairman driven' (1 2) variable for successful exploitative innovation and was conveyed by Respondent A, who commented that:

> The mission statement came from our desire from our chairman, and directors at the time to establish to what [ArchSME] was and where it was going, so it came from senior management. (Innovation 1: mission statement)

Conversely, the senior management were not sufficiently motivated to drive the Learndirect project (innovation 7) into the company. This lack of senior management support was a significant contributory reason for its failure. It was evident in the 'chairman not committed' (7 20) variable and was illustrated by Respondent E, who noted that:

> [The Chairman] gives me the ok [but no more]. You're allowed to do [the Learndirect project], you can run the project. We had the open day, had lots of people attend it and that's about it. (Innovation 7: Learndirect project)

Employee participation

The employee participation in decision-making was seen to be important in successful and unsuccessful exploitative innovation. To make staff feel be part of the development of innovation was seen to be critical to the level of staff motivation to ensure its success. This imperative was epitomised by Respondent D, for example, who stated:

> People get motivated when they are a part of a development, and everybody in the office was made to feel a part of the discussion . . . because to be a part of it . . . then the motivation comes with us. (Innovation 1: mission statement)

It was found that high level of employee participation enabled the knowledge sharing between staff. This was demonstrated by Respondent D, who stated:

> Make people feel a part of the groups and the way you get people to talk . . . share what they thinking is by informal meetings. (Innovation 1: mission statement)

When there was not broad-based ownership of an issue, employees become alienated from the process and 'employees not buy in' (7 11) which resulted in exploitative innovation failure and was evident in Respondent E, who stated that:

> People just don't want to do it. People didn't buy into it . . . . They couldn't be bothered. (Innovation 7: Learndirect project)

However, 'not all employees bought into' (1 6; 2 7) resulted in exploitative innovations which were subsequently successful. Respondent E, for example, described that:

> A lot of people thought [IiP] was another fad. (Innovation 2: IiP)

The distinguishing dynamic in the IiP (innovation 2) was that it was client driven and engaged in significant and enduring senior management championing and day-to-day commitment to its development and implementation.

A supporting mechanism to encourage the appropriate buy in of staff to participate in exploitative innovation, from an HC perspective, was identified as training. This was shown in the 'training' (1 22; 2 9), 'some staff sent to attend training' (5 16) and 'employees encouraged to attend seminars' (5 18) variables. The use of the training to 'raise employee awareness' (1 5; 2 5) was emphasised by Respondent D, who commented that:

> Our industry is based on training. You don't arrive with knowledge; you gain it from this industry. You learn from other companies.... There is a process to sharing knowledge. (Innovation 1: mission statement)

Further, Respondent B explained that the training was used to develop professional knowledge:

> The only thing you can manage the knowledge from is to go on courses. (Innovation 5: seminars)

The firm commitment to training was further confirmed by Respondent E, who articulated that:

> We encourage [employees] to develop themselves...we invest in them with time and money. (Innovation 7: Learndirect project)

Employees are also provided with the necessary finances to participate in external training which they feel will extend and develop their knowledge. This was noted by Respondent B, who said that:

> You are encouraged to attend external courses that you want to do, then you are encouraged to attend it, and then the company will pay the bill for that. (Innovation 2: IiP)

In contrast, two supporting mechanisms concerning the appropriate 'buy in' of staff to participate in unsuccessful exploitative innovation were identified as inappropriate encouragement and the innovation not being related to individual's jobs.

Taking the first issue, 'inappropriate encouragement' was captured in activity including the 'encouragement from all management' (5 34), 'encouragement from top management' (5 28), 'encouragement from team leader' (7 7) and 'encouragement by using the free course' (7 19) variables. Respondent A, for example, described that:

> The support to [attend seminars] is initially committed and encouraged. There is nothing about it specifically but it was encouraged. (Innovation 5: seminars)

It was found that 'encouragement' could sometimes be 'coercive' in nature. Respondent A, for example, stated that:

> We do actually threaten staff...we pay the tuition fees, if you fail to attend these courses on a regular basis, then we have suggested that we may stop paying the tuition fees. (Innovation 5: seminars)

Second, it was seen that motivation of 'buy in' of staff used in unsuccessful exploitative innovation was socially derived which was not transformed to meet project-specific needs. This was evident in 'not related to the job' (5 10) and 'the team not motivated' (5 17) variables. The decoupling between the seminars and individual jobs was captured by Respondent B, who stressed that:

> IT session just related to individuals, not related to jobs. (Innovation 5: seminars)

The team not being motivated was also emphasised by Respondent B, who noted that:

> In terms of motivation, I don't think [the seminars have] motivated the team in anyway. (Innovation 5: seminars)

Lack of time

The notion of 'no one had time' (5 9) was a commonly cited factor in unsuccessful exploitative innovation. The tension between the time and volatility of workload was stressed by Respondent A, stating that:

> ...[the seminars are] purely a failure of whoever was in charge of organising...Something, first of all, you don't have time to do it. Secondly, you have pressures from clients to do the work. It's very difficult to set up the time to deal with the scope we have discussed the project we are working on. The pressures of work removed our ability to handle these sessions. (Innovation 5: seminars)

Similarly, the nature and volatility of workload was expressed by Respondent D, who said that:

> We should look back and said, right, we should do some that; we should do this or we shouldn't do that, and then set it. Something we know we can do because the system is in place. It has the information. We just need the time to look at the information within the team. (Innovation 5: seminars)

In summary, HC was found to be embedded within the capacity, ability and motivation of senior management and employee participation in decision-making. The lack of time was found to be a key obstacle to successful exploitative innovation.

The key distinction between successful and unsuccessful exploitative innovations, from an HC perspective, was the motivation of senior management to drive the innovation through to successful implementation, and to encourage appropriate employee participation in the process.

In successful exploitative innovation, the motivation of senior management to implement the innovation came from top management support. The 'buying in' of staff was encouraged through training which met the unique needs of the teams and individuals. In contrast, in unsuccessful exploitative innovation, top management often did not sustainably commit to the innovation. As a consequence, senior management did not carry out the innovation implementation activities. The staff 'buy in' process was limited to socially derived motivation which was not transformed to meet the unique needs of individuals' roles and project tasks.

5.6.2   Structure capital

The SC for exploitative innovation was found to be principally located in the administrative system, the team structure and computer systems. There were no quantitative innovation performance measurement systems.

The company administrative system took two key forms: appropriate structure and appropriate documentation. First, the importance of an appropriate structure was particularly pertinent in exploitative innovation. The success of exploitative innovation was seen to depend on the formalised structure which was captured in the 'management meeting' (1 26; 5 11) and 'quarterly office meeting' (1 23; 4 6) variables. The acceptance of the innovation was decided by the management board. This was demonstrated by Respondent D, who stated that:

> [Senior management] will have the meeting once a week for senior management, and then they will go back to that team and share that information with the rest of the team. So the process goes through that way. (Innovation 5: seminars)

The quarterly office meeting was used to enable the interaction between different levels, and was captured by Respondent D, who stressed that:

> Initially it was done through quarterly meetings of the whole office.... The process or the structure is laid down by senior management at that meeting. This is what we are doing.... So we're really getting everybody involved and letting them know what is happening through the quarterly meeting. (Innovation 1: mission statement)

By contrast, the 'no structure' (5 12) variable played a crucial role in unsuccessful exploitative innovation. Respondent E, for example, indicated that:

> [The Learndirect project] failed because there is no structure. (Innovation 7: Learndirect project)

The necessity of the formalised structure was evidenced by Respondent A, who stated that:

> We tend to find that if the project is interesting then people will attend. We hold it in the office. We don't hold it in the meeting room. So that is how it stops work anyway. I think the way we make it go forward it is to establish a basic formula every month...a system which is carried out when an interesting project comes in. (Innovation 5: seminars)

The need of a formal structure for the Learndirect project (innovation 7) was demonstrated by Respondent E, who noted that:

> I think we will have to get the structure into [the Learndirect project]. Yeah, structure definitely. Formalise it. (Innovation 7: Learndirect project)

Second, the importance of 'appropriate documentation' was demonstrated in the 'mission statement information documented' (1 7) and 'IiP information documented' (2 16) variables. Appropriate documentation to 'raise employee awareness' (1 5; 2 5) was particularly addressed in successful exploitative innovation. Respondent C, for example, emphasised the relative importance of codifying knowledge in documentary form:

> There are copies of the mission statement document all around the office. We certainly know what it is! (Innovation 1: mission statement)

Respondent E, for instance, explained that managerial efforts were made to ensure that knowledge sharing process happened:

> [Business Development] attached a tick form on the front [of the IiP information] to make sure they ticked their name off and passed it on and make sure everyone had read it. (Innovation 2: IiP)

In contrast, 'nothing recorded' (5 30) was stressed in unsuccessful exploitative innovation, with the impact that the issue and lesson learned could not be encoded and documented. This was evident in 'good ideas not captured' (5 24) variable and was captured by Respondent A in the case of seminars, who said:

> Nothing was recorded because it's informal. (Innovation 5: seminars)

Specifically, a lack of time (discussed in Section 5.6.1) to take the minutes of seminars was emphasised by Respondent D, who stated that:

> [The seminars are] more informal. That is, it isn't really minuted or reports done or anything. That's just more time. (Innovation 5: seminars)

Appropriate documentation was seen as the key mechanism to reinforce exploitative innovation. This was evidenced by Respondent A, who stated that:

> We started doing an attendance record. It sounds high and almighty, but it is the way to make sure people will turn up. If you don't turn up, if you haven't given a good excuse it will be noticed. (Innovation 5: seminars)

Although 'everyone had a PDP' (7 17), the Learndirect project (innovation 7) still failed. This failure was found to be caused by the role of senior management (discussed in Section 5.6.1).

In combination, these variables show that the formalised system with appropriate structure and documentation within SC was critical for successful exploitative innovation.

In successful and unsuccessful exploitative innovations, it was found that there was 'no specific way to measure the innovation performance' (1 21) within ArchSME. There were no formalised measurement systems; rather, there were mechanisms such as 'annual staff appraisal' (2 10) and 'informal meeting' (7 5), but they did not explicitly or adequately address this issue. The determination of the perceived success or failure of an innovation was through informal, daily feedback, expressed by Respondent E, who noted that:

> I have a chart to measuring people progress, but its not really measuring it in that kind of way. I just keep an eye on them. (Innovation 7: Learndirect project)

When it comes to feedback, the only formal feedback system for learning was the annual staff appraisal. Evaluations are often annual and were therefore regarded as a slow, if not irrelevant, feedback system.

The SC for exploitative innovation was supported by an enabling 'team structure'. The importance of stimulating and developing teamwork at an operational level was evident in the 'informal team meeting/discussion' (4 8; 5 6) variable which was raised by Respondent D, who stated that:

> For something to be supported it, it needs to be shared. So we have, we share with the team, the whole team discuss it. (Innovation 4: company restructure)

Within the team structure, the key distinction between successful and unsuccessful exploitative innovations, from an SC aspect, was that successful exploitative innovation was characterised by enduring senior management support from inception through to implementation (discussed in Section 5.6.1). The importance of 'senior management implementation' (1 1; 2 4; 4 5) was seen to be essential in successful exploitative innovation. This was described by Respondent E, who stated that:

> We didn't really consult [our staff]...because [IiP] was more about the processes and things like that that top management had to put it in place. It didn't really involve our staff much because apart from getting them to buy in, there wasn't really much else to do. (Innovation 2: IiP)

In unsuccessful exploitative innovation, it was found that senior management did not drive the implementation through the team structure. The support from management level in innovation activity, including 'senior management chose attendees' (5 5), 'senior management led it' (5 19), 'middle management implementation' (5 8), 'business management led it' (7 21) and 'business management monitored the progress' (7,18) variables. Respondent A, for example, commented that:

> the failures all come from the management. (Innovation 5: seminars)

Lack of senior management endeavour to drive the innovation into the organisation, resulted in exploitative innovation failing. This was evident in the 'management not drive it' (5 4) and 'senior management not drive it' (7 12) variables and was emphasised by Respondent E, who stated that:

> I got [senior management] commitment, but they didn't drive it down the organisation. (Innovation 7: Learndirect project)

The need of 'senior management' to drive the innovation into the organisation was emphasised by Respondent E, who stated that:

> I suppose in the next year, when we come back from Christmas, I will get the senior management to drive [the Learndirect project]. That will make a big difference. (Innovation 7: Learndirect project)

In a computer system context, SC took two key forms: the computing programme and the company website. The 'company website' (1 11; 5 27) was seen as a significant activity in supporting exploitative innovation. Respondent A, for example, explained the importance of the company website:

> The website is the biggest thing that we have done recently to support [the mission statement]. (Innovation 1: mission statement)

The 'computing programme' (5 15) was particularly addressed in supporting exploitative innovation. Respondent A, for example, stressed that:

> . . . something like our job costing programming system, which is not necessarily new to us, but it does very well [it that it] helps me . . . my management. (Innovation 5: seminars)

In summary, the principal locus of SC within exploitative innovation was found to be the formalised administrative system (with appropriate structure and documentation), the team structure and computer systems. There were no quantitative innovation performance measurement systems. Successful exploitative innovation was found to have formalised structures and documentation systems, enduring senior management support from inception through to implementation and supported by an enabling team structure which stimulated and developed teamwork at an operational level. In contrast, unsuccessful exploitative innovation was found to have no formalised structures and documentation systems and no senior management support to drive the innovation down into the organisation.

5.6.3   Relationship capital

The key sources of RC for exploitative innovation were located within business adviser, internal, client and supplier interactions.

At business advisers' interaction level, RC is seen as being important in terms of 'operational' interaction to fulfil the knowledge gap which ArchSME did not have on its own. The 'business advisers' (1 27; 2 15; 7 1; 7 4), 'free resources from government' (7 3), have significant influence in the process of knowledge creation in exploitative innovation. In successful exploitative innovation, the need of the mission statement came from the business adviser and was captured by Respondent E, who stated that:

> [The idea of the mission statement] came through IiP, Investors in People. So it came through [business advisers], they said that if we have the mission, we will have more focus. (Innovation 1: mission statement)

The idea for the unsuccessful Learndirect project exploitative innovation was from the business adviser in an 'informal chat in the open day' (7 9) and was demonstrated by Respondent E, who noted that:

> [The idea of the Learndirect project] came from our business advisers again, consultants. (Innovation 7: Learndirect project)

Interaction between knowledge workers and colleagues was emphasised in the 'good relationship with colleagues and suppliers' (1 10) variable. RC within an internal context through team structure at 'operational level' and 'social level' interactions was evident. At an operational level, the 'informal team meeting/discussion' (4 8; 5 6) was emphasised by Respondent D, who stated that:

> A lot is done informally. Talking again. From talking to our client to look our portfolio because that is really our business we are showing with the portfolio. [The team] will then talk to them about our company which is what we are aiming for, which is what we do. (Innovation 4: company restructure)

Knowledge workers and colleagues interactions at a social level were captured in activity including 'informal discussion/meeting in the office/pub' (1 14; 1 18; 2 12; 5 2), 'informal meeting' (7 5) and 'social activity' (1 16) variables. In successful exploitative innovation, this was demonstrated by Respondent C, who noted that:

> Sometimes we will go out, say, and play football together with sometime from a different team who works on a different floor who I don't see on a daily basis. Sometimes the company goes out, the whole company. (Innovation 1: mission statement)

In unsuccessful exploitative innovation, the social level interaction through 'the team structure' was emphasised by Respondent D, who explained that:

> We have that interaction on that level with the whole company … the different [teams] interact at a social level. (Innovation 5: seminars)

At a client interaction level, RC is viewed as being important in terms of 'operational' interaction to progress specific project issues, and to establish a foundation for the company marketing.

In successful exploitative innovation, the client was identified as being the principal operational RC focus. This was evident in 'the client wanted to know all team members' (4 2) variable and was described by Respondent D, who commented that:

> A lot of clients … like to know all members of the team. When they pick up the phone they who they are speaking to. They know that they can come back to the same person. So we don't just deal with senior management. We need to deal with each level because they are the people drawing the information. They are the one who has the most knowledge. Therefore, they can share it. So, but they need to understand who draws within the team, the people. (Innovation 4: company restructure)

By contrast, 'the client's job has higher priority' (5 22) over non-client activity was a significant contributory reason for exploitative innovation's failure. This view was described by Respondent D, who expressed that:

> Other things come in which have a higher priority, primarily because we are still in the commercial business and if the work needs to be done and then it needs done. The client cannot wait because we have internal meetings. (Innovation 5: seminars)

It was found that marketing within ArchSME is very much enmeshed with identifying and understanding particular clients, and this process was found to be proactive and informal in nature. Respondent A, for example, stated that:

> The marketing within the company is very informal and involves entertaining clients really. (Innovation 1: mission statement)

The informal nature of marketing was reinforced by Respondent A, who claimed that:

> A lot of jobs are through the words of mouth. The informal marketing is very important. (Innovation 1: mission statement)

Interaction between knowledge workers and suppliers was emphasised in the 'good relationship with colleagues and suppliers' (1 10) variable at an 'operational' level. Respondent C, stated that:

> We have the good relationship with other professionals we use on a regular basis, other consultants. (Innovation 1: mission statement)

In summary, RC for exploitative innovation was located at business adviser, internal, client and supplier interaction domains of activity. RC seems particularly crucial to knowledge creation.

In the cases of successful exploitative innovation, it was found that 'operational' and 'social' RC sources fed into specific project needs. In contrast, the unsuccessful exploitative innovation was underpinned solely by 'social' RC sources which did not meet project-specific innovation needs, such as internal organisation and general client development activity.

5.6.4   Knowledge capital

The KC for exploitative innovation was associated with a combination of 'social' and 'technical' contexts. In a 'social' context, innovation activity was seen to take place in the company environment (such as office and open family culture) and the public house. This was shown in the 'informal discussions/meetings in the pub/office' (1 14; 1 18; 2 12; 5 2), 'office' (1 17) and 'open family culture' (1 28; 2 3) variables. The company environment in ArchSME serves as an important symbol of professionalism. The importance of the office to gather people together and to 'raise employee awareness' (1 5; 2 5) was captured by Respondent C, who stated that:

> The office has a quite good social structure as well. Lot of people come together and play football, and structured nights out with the company, curry night, and things like that, good for team building, that kind of thing. (Innovation 1: mission statement)

The open 'family' culture was particularly addressed in successful exploitative innovation. Respondent E, for example, illustrated that open family culture enabled employees to work towards a common goal:

> [Supporting IiP] really comes from the open family culture again. Supported investment in people. We had good employee buy in for it . . . they could see the benefits for themselves as well as for the business. (Innovation 2: IiP)

A 'technical' context was seen to complement to a 'social' context. The object of exploitative innovation was found to be the generation of organisation-wide SC. Two types of mechanisms were used in a technical context: emails and internet searches. The use of 'emails' (1 8; 2 1; 5 29; 7 6) to 'raise employee awareness' (1 5; 2 5) was demonstrated by Respondent B, who noted that:

> Like an email which lets you know what is going on in the company. (Innovation 2: IiP)

The use of the 'internet' (7 10) was particularly stressed in the unsuccessful exploitative innovation implementation phase. Respondent E, for example, stated that the Learndirect project was an online training:

> They have the open day. The learning is done through [business advisers'] company on the website. (Innovation 7: Learndirect project)

In summary, KC for exploitative innovation was associated with a combination of 'social' and 'technical' contexts (see Section 5.5.4 for the description of the social and technical contexts). First, in a 'social' context, KC stimulated interaction and collective 'process-orientated' knowledge creation and conversion. This took the form of office environments which supported team activity, such as meeting rooms and office layout. Second, in a 'technical' context, KC supported the capture, storage and retrieval of 'asset-orientated' knowledge. This took the form of emails and the internet searches.

The key distinction between successful and unsuccessful exploitative innovations was the sources of ideas and their application. In successful exploitative innovation, KC was used to channel to project-specific, operational activity. In contrast, in unsuccessful exploitative innovation, KC did not meet specific project needs (for example, online training for individual needs).

5.6.5   Innovation outcomes

The outcome of exploitative innovation was found to improve organisational effectiveness and efficiency. It was evident in the 'improved business performance' (2 17) variable and was captured by Respondent E, who noted that:

> Improve business and then again retention, recruitment and attraction, and turnover. (Innovation 2: IiP)

The positive outcomes from exploitative innovation were reflected in five aspects in organisational performance: strategic direction, formalised structure and process, team-based performance measurement system, staff motivation and recruitment, and company marketing.

Strategic direction

The outcome of exploitative innovation was found to give the company strategic direction and was demonstrated by 'company had future direction' (1 15; 2 8) and 'improved company confidence' (2 13). For example, the use of IiP as company strategic direction was emphasised by Respondent E, who stressed that:

> ...we use the IiP as a spring board, to do different things like EFQM [European Foundation for Quality Management]. (Innovation 2: IiP)

The formalised structure and process

The introduction of a formal structure and process through implementing exploitative innovation has improved the process effectiveness. It was evident in the 'company had structure and process' (4 9), 'company had structure' (1 14) and 'company had process' (2 11) variables. Respondent B, for example, stated that some standard procedures were established:

> We started to carry out the process we previously wouldn't have thought about when we were small. (Innovation 2: IiP)

Team-based performance measurement system

The introduction of the company structure helped the management in evaluating the performance of teams. It was present in the 'company had team-based measurement system' (4 11) variable and was captured by Respondent D, who stressed that:

> We are able to look at that team. The director can just look at, to address, that team, that's say, how much time within that team has been spent and what has been done by that team rather than look at the whole company, he can just look at that specific team and he is able to do that we the systems that we have, and then they come back to the team leaders, and they look at that... is there any issues, and then they go from there. (Innovation 4: company restructure)

Staff motivation and recruitment

The outcome of exploitative innovation was seen not only to encourage the retention of staff but also to attract people to join the firm. The staff motivation and recruitment was demonstrated in the 'motivated staff' (1 25) and 'recruited new staff' (1 24) variables. Respondent A, for example, indicated the use of mission statement to contribute to the socialisation of new staff:

> We use it...to achieve, to gain staff. The staff we give we have to buy into the mission statement maybe mindset.... So that staff may be will be attracted in the mission statement. (Innovation 1: mission statement)

It was found that the motivation came from 'staff understood the firm more' (1 12; 5 23) and 'clients and staff understood the firm more' (4 10). Respondent A, for example, stated that the staff is motivated by the mission statement:

> The staff needs to be motivated. I think I cannot see the mission statement motivates people, but I think it gives more understanding of the firm. If you get more

understanding of the firm, how it's being run, then you feel your belong or by that effect you should feel more motivated. (Innovation 1: mission statement)

Company marketing

The importance of 'badges' was seen as important marketing devices. The appearance of IiP was crucial for market reputation; and, the burden of maintaining the emphasis of 'people focus' reputation was something that both senior management and knowledge workers collaborated in sustaining. This was evident in the 'improved the company reputation' (2 14) variable and was emphasised by Respondent B, who commented that:

> The company name seems to be known a lot more. (Innovation 2: IiP)

This enhanced reputation was felt to be important in attracting the company's major clients and new clients. Respondent D, for example, indicated that the company's major clients had an interest to know the company's mission statement:

> [The mission statement] matters to some clients more than others. Some organisations they look at the mission statement; they would expect us to have a mission statement and feedback to the company they know where we are going. With others not interested. They want to see the work – not this! Yes, there is a benefit for some major clients – we know where we want to go. (Innovation 1: mission statement)

The benefit for identifying the company itself was evident in the 'company had identity' (1 13) variable and was confirmed by Respondent A, who stressed that:

> [The mission statement] defines our products; it explains how our management is working and who our products are working for, and also it gives the company identity which we never had. (Innovation 1: mission statement)

As a consequence, this identity could be 'used in the marketing' (1 20). For example, Respondent E expressed that the company used the mission statement in tendering and the marketing:

> We used our mission statement when we wrote our tenders and bids, so we advertised it and put it on the website as well. (Innovation 1: mission statement)

It was found that unsuccessful exploitative innovation also contributed some unexpected benefits at an individual level. It was evident in the 'discovered some staffs' other skills' (5 1), 'staff engaged in some projects more' (5 21), 'increased knowledge' (5 35) and 'some staff learned some skills' (7 13) variables. Respondent A, for example, stated that:

> Discovering that within some of our teams, some of the younger architects or technicians were quite good in presenting and also gained confidence in presenting in front of staff. (Innovation 5: seminars)

Nevertheless, the outcome of exploitative innovation proved to erode organisational performance. The negative impacts from exploitative innovation

Table 5.2 Variables in explorative and exploitative innovations.

Type of innovation	Variables	Generic variables	Distinctive variables for successful innovation	Distinctive variables for unsuccessful innovation
Mode 1: Explorative innovation	HC	The capacity, ability and motivation of staff	Social or operational knowledge being applied to meet the project needs	Social knowledge not being applied to meet the project needs
	SC	Team structure Teamwork	Team-based ideas Teamwork Senior management involvement through teamwork	Individual-based ideas Individual-based work Senior management not involved in teamwork Limitation of relevant and updated information within the structure
	RC	Operational RC: within internal, client and supplier interactions Social RC: within internal, client and supplier interactions	A combination of operational RC and social RC being applied to meet project needs	Social RC not being applied to meet project needs
	KC	Social context: company environments (office and meeting room) Technical context: emails and the internet	A combination of social context and technical context	Technical context
	Outcome	Effective and efficient delivery of services to satisfy current and/or future project needs	Project performance improvement	Individual performance improvement

Mode 2: Exploitative innovation				
HC	The capacity, ability and motivation of senior management Employee participation	Top management support Senior management implementation Some employees buy in Training	Top management not supportive Senior management not driving the implementation Lack of time Employees not bought in Inappropriate encouragement Not related to an individual job
SC	The administrative system Team structure Computer systems	Formalised structures and documentation systems Senior management implementation through the team structure	No formalised structures and documentation systems Senior management not driving the implementation through the team structure
RC	Operational RC: within business adviser, internal, client and supplier interactions Social RC: within internal interactions	A combination of operational RC and social RC being applied to meet project needs	Social RC not being applied to meet project needs
KC	Social context: company environments (office and open family culture) Technical context: emails and the internet	A combination of social context and technical context being applied to meet the project needs	A combination of social context and technical context being applied to meet the project needs
Outcome	Organisational effectiveness and efficiency	Organisational performance improvement	Individual performance improvement

HC, Human capital; SC, Structure capital; RC, Relationship capital; KC, Knowledge capital.

were demonstrated in the 'too much work' (4 12), 'took too much time' (5 20), 'it's stopped' (5 32), 'cost a lot of money' (7 14) and 'lost training opportunity' (7 15) variables. Respondent D, for example, complained the unbalanced workload between teams:

> Balancing sometimes. Amount of work we do within the teams.... Sometimes, the work is too much. (Innovation 4: company restructure)

Respondent E mentioned that 'something wrong with company IiP' (7 16):

> [The Learndirect project] ties in with IiP...if we're failing with that then we've obviously done something wrong with IiP. (Innovation 7: Learndirect project)

In summary, exploitative innovation was seen to improve organisational effectiveness and efficiency, and generate sustainable competitive advantage. The successful exploitative innovation was seen to improve organisational performance. In contrast, unsuccessful exploitative innovation was found to improve only individual performance, rather than collective, organisational performance.

5.7   Summary and Link

This chapter has presented the key findings from the exploratory phase of the case study. Two types of innovation in ArchSME were identified: exploitative innovation and explorative innovation. Key variables around company innovations are summarised in Table 5.2. These variables, and their interaction, were further explored and tested in the action research phase set out in the next chapter.

6

Case Study – Action Research Phase

6.1 Introduction

This chapter describes the key results from the action research phase of the case study, and in so doing, further tests and develops the findings from the exploratory phase. This chapter is structured using the action research cycle phases discussed in Section 4.3.4. Each phase is divided into two sections. First, the 'practice' undertaken in the action research is described. Second, the researcher's 'reflection' on that practice is presented. This discussion is structured using human capital, structure capital, relationship capital and knowledge capital variables (see Table 5.2).

6.2 Diagnosis

6.2.1 Practice

The 'start' of the diagnosis phase was a company workshop. Its purpose was to discuss and evaluate the key findings from the exploratory phase (see Chapter 5), and based on this, to identify an action research intervention or innovation to be developed and implemented. The workshop members consisted of seven participants: five ArchSME respondents involved in the exploratory interviews and two members from the University of Salford.

There were two main stages in the workshop (see Section 4.4.4). First, the researcher presented the key findings from the exploratory phase. This stage was designed to stimulate a discussion by the group with a set of questions given in the company general finding report (see Appendix E). The main sections of the report took the form of questions. These were as follows.

The first question was 'what are the immediate innovations which ArchSME should progress?' Two potential innovations, along with their objectives, benefits and resource implications, were listed. There were an exit planning and a post-project review protocol.

The second question was 'what is ArchSME's current position?' It was found that the company was good at external innovation (explorative innovation) to solve one-off client problems, but not so good at internal innovation (exploitative innovation) to improve operational efficiency.

The third question was 'what are ArchSME's potential problems?' This question was divided into two sub-questions. In the first sub-question, ArchSME's current position was discussed. In the second sub-question, ArchSME's potential problems were articulated. It was found that with the increasing growth of the firm, the limitation of current internal systems will probably become a restraining force.

The fourth question was 'why manage knowledge?' Based on ArchSME's respondents' perspective, there were five sub-questions under this main question. In the first sub-question, 'what is knowledge?' was introduced. The second sub-question addressed the question 'where knowledge is?' The third sub-question illustrated 'what is knowledge management?' The fourth sub-question expressed 'why manage knowledge?' The final sub-question was 'what are the potential benefits of managing knowledge?'

The fifth question was 'what are potential improvement areas to sustain current growth?' The potential improvement areas for ArchSME were identified under the following classification: immediate wins, short-term wins and mid- to long-term wins.

The final question was 'what are the key findings?' This section was a summary of the above questions.

The ArchSME representatives found the results of the company general finding report (see Appendix E) interesting and valid. Respondent E, for example, gave the feedback as:

> The presentation looks great! It gives some good practical examples too.

The second stage of the workshop moved on from the general debate to focus on the two proposed immediate innovations – exit planning (exist interview) and a post-project review process. Both potential innovations were stressed in the exploratory phase as being high priority issues to be addressed. The first proposed immediate innovation was exit planning. During the exploratory phase of the case study, it was found that there was no procedure in dealing with employees leaving the practice. The exit planning innovation was expected to capture and share important knowledge from staff leaving the practice, and to ensure stability and continuation of client service when key staff leave. The second potential innovation was a post-project review process. ArchSME did not have any procedures to learn from project activity and measure project performance. Further, the company lacked appropriate structure and communication channels to encourage and support knowledge transfer between 'ring-fenced' project teams in a formal way. Respondent D, for example, described the benefit of having a post-project review in the company system:

> ... if we did [post project reviews], then it would save time in the future and money from repeating mistakes.... We should, but we don't really have it.

The post-project review process innovation was expected to identify areas for improvements, reduce employees 'reinventing the wheel' or repeat their mistakes in future projects and help to build a stronger sense of commitment and team spirit.

Table 6.1 Company workshop minutes.

Project name					Innovation research and development project	
Date	15 May 2004 (Thursday)	Duration	12:00 to 13:15	Venue	ArchSME meeting room	

1. **Key project issues**
   - The 'interim project review' has been chosen as the company innovation.
   - It is proposed that the deliverables of this project will be the interim project review policy, guidelines and checklists, and then will be integrated into the ISO 9001 quality management system.
   - The reviewer should be the 'the architect' rather than 'the project leader' or 'the associate director'.
   - The client will be involved in this project. Thus, there is a need to define the role of the client and what benefits will be provided for the client.
   - The company will identify a project and a task group to cooperate with the Salford researcher in conducting this project.
   - The interim project review policy, guidelines and checklists should be tested by all teams.
2. **Responsibility**
   - The Salford researcher will work in the company and provide own laptop (from 24 May to 23 July).
   - ArchSME task group leader will be responsible for allocation of staff to engage in this project, for example, arranging the meetings, etc.

The adopted innovation was thus to develop and implement an interim (rather than post) project review process into the company. The rationale for this prioritisation was that ArchSME did not have any systems of this in kind in place with, as an inevitable result, good practice, and lessons learned were not being captured and shared for future projects. At the time, ArchSME was preparing for ISO 9001 quality management system accreditation.

The associate director championed the innovation. He expressed support in providing appropriate access for the researcher to be involved in the development and implementation of the action research intervention, and for allocating ArchSME staff to form a task group. The group consisted of the researcher from the University of Salford and staff from ArchSME. The role of a task group was to develop and implement the action research intervention. The ArchSME quality representative was the leader of the task group.

The company workshop minutes are shown in Table 6.1. There are two sections involved in the minutes. First, the *object and the key issues of this project* section clarified the key issues raised in the workshop and recommended issues of action. Second, the *responsibility* section identified the roles and responsibilities of the researcher and the ArchSME staff in the task group.

6.2.2 Reflection

The adopted innovation – interim project review process innovation – was categorised by the researcher as an exploitative innovation as it focused on an internal organisation process which was not being developed for a specific project (see Section 5.4). The key variables for exploitative innovation were discussed in Section 5.6 and summarised in Table 5.2. The discussion in this section is structured around the human capital, structure capital, relationship capital and knowledge capital variables.

Human capital

The two generic variables within human capital for exploitative innovation identified in the exploratory phase were the capacity, ability and motivation of senior management; and employee participation. The distinctive variables between successful and unsuccessful exploitative innovation, from a human capital perspective, were top management support, senior management implementation through the team structure, buy in of employee, and the need of time to develop and implement the innovation activity (see Section 5.6.1). The principal variable at work in the diagnosis phase of innovation appeared to be the 'senior management' role.

The discussion during the workshop reinforced 'the capacity, ability and motivation of senior management' variable. The debate was principally led by Participant A (a senior manager) and Participant D (a team leader) from ArchSME, and the two researchers from the University of Salford. The other three, more junior, participants from ArchSME appeared unwilling and/or unable to shape the flow of the discussion.

With respect to the first of the two proposed innovations, Participant A disagreed there was a need for an 'exit planning procedure' due to the low rate of employee retention:

> 90% of staff has remained with us throughout [since the formation of ArchSME in 1991].

This opinion was not challenged by the other ArchSME delegates. The discussion then moved to the second proposed exploitative innovation - post-project review. This idea was contested by Participant A, who commented that:

> I don't think you can abstract that huge information from [the post project review].

Participant D, however, disagreed with his view and suggested that there was a need of a 'post-project review':

> We are learning from each project – where we will spend time, where we will spend money. We should, but we don't. We should assess at the end of each project within the team. We should assess what went wrong and why, and why we don't do it. Primarily we don't have time to do it. So we hope in the future we should be developing systems to assess how we can better be able to do things or learn from other things.

Participant A modified his view based on this agreement, and advocated that:

> Sometime, obviously, knowledge is difficult to tap it into within the practice.... The project review system might help.

Participant D supported his view and asserted that:

> The project is not about three or four weeks. It's about three or four years.

In response to this, the idea of an interim project review was stressed by Participant A, who stated that:

> ... an interim project view on how [the project] is running would be useful.

The idea of an interim project review process as the focus of the action research phase was supported by Participant A. This intervention was prioritised by the associate director (senior management). This is consistent with the key findings from the exploratory phase which emphasised the pivotal role of senior management in exploitative innovation. Further, it was found that senior management has a significant impact on engendering enthusiasm for new ideas amongst staff. After the associate director committed to the interim project review process innovation, other participants showed their 'high' interest to be involved in this project. Participant C, for example, stated that:

> Yes, I think the interim project review is a great idea.

This indicates that the initial level of employee participation was influenced by senior management.

In summary, the key role of senior management in framing and prioritising innovation activity within the diagnosis phase was confirmed.

Structure capital

When considering the structure capital aspect, the administrative system, the team structure and computer systems were found to be the generic variables in exploitative innovation. The distinctive variables for successful exploitative innovation were the presence of formalised structures and documentation systems, and senior management endeavour to drive the implementation through the team structure (see Section 5.6.2).

First, the need of a 'formalised structure' into the interim project review process was immediately captured by Participant A, who noted that:

> ... the idea must be formalised into the process. I don't know how we do that.

The argument for formalisation was counterbalanced with a need to keep any process 'resource light,' and to be sympathetic to current work practices. This argument was advanced by Participant D, who stressed that:

> From my point of view, do we actually want to go down the Investors in People path? That's formal. Sometimes we need to stay informal. That's the way we learn, trying to demonstrate in, it's not just detail, but contact .... The review comes from a couple of people sitting in [ArchSME] and knowing what somebody is doing. That's not something necessarily to formalise into charts or client satisfaction etc. It's sharing knowledge and ... how you reuse that information. So I think [the project review] will fix this.

These arguably opposing views of 'formal' versus 'informal' were resolved by Participant A, emphasising the need to:

Make this review activity easy, simple and manageable.

Participant A then fixed the responsibility and authority for the review at the architect level:

... probably the architect to do the review rather than the associate director or the team leader to do the review.

Second, the need of the team structure to implement the interim project review was noted. The researcher and the ArchSME quality representative led the development and implementation of the interim project review process innovation, with the associate director being the senior management champion.

In summary, the key role of the formalised structures and documentation systems and the key role of senior management endeavour in driving the innovation implementation through the team structure within the diagnosis phase were confirmed.

Relationship capital

The two generic variables within relationship capital for exploitative innovation in the exploratory phase of the case study were operational relationship capital and social relationship capital. The key distinctive variables between successful and unsuccessful exploitative innovations were the sources of the ideas and their application, i.e. for a specific project or for general organisation capability (see Section 5.6.3).

The issue of encouraging client involvement in the development of the interim project review process innovation was advanced by Participant A, who stated that:

The more I get interested in this, I want to get the client involved [in the interim project review process].

Participant A stressed the benefits of such client involvement, in the observation that:

Learning back from the previous successful project, the more important it is to develop in more depth the relationship with clients.

The proposed interim project review process innovation addressed the need to more adequately capture feedback from the client, both within respect to the 'content' of the work being delivered to the client and the 'process' of how it was being delivered. The opportunity to further develop deeper relationships with clients was addressed by senior management. This 'opening up' of the internal workings of the firm to the client was perceived as being a stimulus for ongoing internal innovation and project-to-project learning, supporting the closer mutual development and successful delivery of the client brief and the forging of deeper, 'whole firm' relationships with clients (i.e. not just between firm associate directors and clients, but with technicians, and so on). This stressed the importance of clients and internal interactions at an 'operational level.' The interim project review process development, however, was not targeted at a specific 'live' project; rather, it was envisaged that the new process would be part of the general organisational endeavour to gain ISO 9001 accreditation.

In summary, relationship capital in the diagnosis phase was located at a social level. The interim project review process innovation (exploitative innovation) was targeted at internal organisation activity, but not at a specific project. This is consistent with the key findings from the exploratory phase.

Knowledge capital

The knowledge capital for exploitative innovation was the focal or integrating nexus in which innovation takes place in social and technical contexts. The distinctive variable for successful and unsuccessful exploitative innovation was that knowledge capital was channelled to for a specific project or for general organisation capability (see Section 5.6.4).

In a social context, the company workshop in the boardroom encouraged face-to-face discussion and sense-making. There was no client or supply chain relationship capital engagement. In a technical context, two mechanisms were used. First, 'the company general finding report' provided the clear aims and objectives for this workshop. Second, 'email' was the main technical tool used to enable communication between the researcher and the main contact person (Participant E). The interim project review process, however, did not target a specific project; rather, it provided a generic process to assist in ISO 9001 accreditation.

In summary, knowledge capital in the diagnosis phase was initially stimulated through the 'technical system' through the company finding report and by communication via email. This provided the platform to commit ArchSME staff to the 'social system' workshop. The source of the ideas and their application was to improve general organisation capability. This is consistent with the key findings from the exploratory phase.

6.3   Action Planning

6.3.1   Practice

Activity 1: development of interim project review action plan

After the company workshop, the documents related to ArchSME's ISO 9001 quality management system were sent to the researcher by its quality representative in May 2004. These documents were produced by ArchSME's external ISO consultant, including the draft of the ArchSME quality manual and its partnership ISO 9001 action plan (see Appendix A). After reviewing these documents, the researcher identified two key issues.

(1) The basic framework for the ArchSME ISO 9001 quality management system was already in place. ArchSME's 'product' in its ISO 9001 system was identified as 'architectural designs and services'. Two broad types of services within the firm were identified as 'traditional contract' and 'design and build contract'.

(2) ArchSME, at that time, did not have any systems or evidence against the ISO 9001: 8.2.3 monitoring and measurement of processes and ISO 9001: 8.2.4 monitoring and measurement of product.

On the basis of the key issues set out in the minutes of the company workshop (see Table 6.1) and the documents which the ArchSME quality representative sent, the initial interim project review process action plan was developed by the researcher (see Table 6.2) and sent to the quality representative in May 2004.

The task force collaboratively developed an action plan for the development and implementation of the interim project review process innovation. The plan was structured around a number of main questions (see Table 6.2), namely what is an interim project review, what is the object of this innovation activity, what is the scope of this interim project review action, what commitment is required from ArchSME, who benefits from the interim project review arena and what is the intervention plan? The initial action plan provided a basis and focus for the collaborative action research activity.

Based on the action plan, the researcher should have started working within ArchSME from May 2004. The researcher decided to arrange a follow-up meeting with the leader of ArchSME task group (the quality representative) to move the innovation forward.

Activity 2: meeting with ArchSME's quality representative

The meeting with the ArchSME quality representative took place in June 2004 in its meeting room. Its objectives were to:

(1) assess the organisation's level of compliance against the ISO 9001 standard;
(2) clarify the delivery of work process in ArchSME;
(3) confirm other members of the task group from ArchSME; and,
(4) confirm the date the researcher could start working within the firm.

With respect to the first issue, there was no difficulty in gaining access to confidential information/documents. These documents related to the ArchSME practice included examples of job forms, drawing issue sheets, site record sheets and so on; and, related ISO documents (see Appendix A). The researcher found that documents related to the ArchSME ISO 9001 system were formalised and documented and were stored electronically. Documents related to the firm's daily routine work, however, were handwritten.

Table 6.2   Action plan: interim project review project.

1.   What is an interim project review?
- An interim project review is an activity where people review what went well and what went badly during the project.
- The aim of this review is to praise each other on jobs well done as well as find ways to do things even better.

2.   What is the objective of this project?
- Develop and test the interim project review policy, guidelines and checklists.
- Help the company to integrate the interim project review activity into the ISO 9001 quality management system: 8.2.3 monitoring and measurement of processes or/and 8.2.4 monitoring and measurement of product.

3.   What is the scope of this project?
- Focus on 'project' level: from establishing feasibility, agreeing design and obtaining permission, supervising traditional contract and overseeing construction (refer to QP4 feasibility and planning, QP5 traditional contract and QP6 design and build contract).

4.   What commitment is required from ArchSME?
- Identify the specific project
- Identify the actors (participants)
  - ☐ The task group (the project team)
  - ☐ The clients (the stakeholders)
- Provide 'space' for the Salford researcher

5.   Who benefits from this project?
- The company level: to improve processes efficiency or/and to ensure that the architectural service provided meets client expectations.
- The client level: **(unknown at this stage)**

6.   Project plan

Table 1 – Action plan for the interim project review

Activity	Method	Duration (May to July)	1	2	3	4	5	6	7	8	9
1 Analyse current practice in more depth 1-1 Identify the role of the actors 1-2 Identify key performance indicators	Access to company documents Interview with the task group	24/05/04 to 04/06/04	▓	▓							
2 Develop pilot policy, guidelines and checklists	Access to company documents Interviews with the task group	07/06/04 to 18/06/04			▓	▓					
3 Review/redefine policy, guidelines and checklists	Interviews with the task group	21/06/04 to 25/06/04					▓				
4 Test (when appropriate) policy, guidelines and checklists cross-teams	Involvement in appropriate company activity	28/06/04 to 09/07/04						▓			
5 Analyse the test results	Use computer software to analyse data	12/07/04 to 16/07/04								▓	
6 Review/redefine policy, guidelines and checklists	Interviews with the task group	19/07/04 to 23/07/04									▓

Time scale (week)

The 'delivering of work' in process was divided into three procedures against the ISO 9001 which are:

(1) feasibility and planning procedure;
(2) supervise traditional contract procedure; and,
(3) oversee construction procedure.

The researcher recognised that there was a need to make the interim project review process fully integrated with the existing ArchSME quality assurance infrastructure. The researcher, however, found that it was very difficult to do so. For example, the researcher found that the firm's procedures confused 'product' and 'process' views, such as the feasibility work being mixed up with the company marketing and the architectural work (traditional contracts and design and build contracts). The ArchSME quality representative, however, could not make a distinction between these three procedures. The quality representative suggested that the interim project review process should cover the whole business process rather than focus on the project level:

> I think [the interim project review process] should cover these three procedures.

The researcher disagreed with this view and pushed through the proposition that the objectives of the interim project review process at the project level, and to integrate it with the ArchSME 'existing' ISO 9001 system (see Figure 5.3 for the description of the commission and delivery of work processes in ArchSME). The researcher found that, for example, the company lacked evidence against ISO 9001: 7.3.1 design and development planning. For instance, the evidence against ISO 9001: 7.6 control of monitoring and measuring devices within quality manual was:

> ISO 9001: 2000 is not relevant and is excluded.

The researcher, however, disagreed with this argument and believed that building regulations, for example, was one of ArchSME's monitoring and measuring devices. This assertion was accepted by the quality representative.

The final two issues – when the researcher could start working within the firm and the allocation of staff to the task group – were not addressed.

6.3.2   Reflection

Human capital

The researcher realised that there were two practical problems with the development and implementation of the interim project review procedure from a human capital perspective. First, there was no ArchSME staff trained and experienced in ISO 9001 quality management systems. Within the action research team, the researcher was the only person with expertise and experience in implementing ISO 9001 within construction companies. The researcher found that there was real difficulty in communicating at an 'expert' level with the ArchSME quality representative. The quality management expertise required for the innovation was largely outside of firm and it had to rely on external sources of capability (in particular, the external ISO consultant).

Second, resources, in the form of time and staff allocation, were still the main constraint in this collaborative endeavour. The initial aspiration was for the action plan to be co-authored by the researcher and the ArchSME quality representative (Participant E). The co-authorship aspiration was aimed to ensure that the plan was appropriate in focus and to assist in creating shared ownership of the interim project review process. However, this co-authorship did not take place, with ArchSME relying solely on the researcher. The sign-off the action plan by the firm's quality representative was done by email as follows:

> Everything is extremely hectic here at present – not had time to think!...The project review proposal is fine.

The researcher found that the leader (the quality representative) of the ArchSME task group did not provide proactive leadership; rather, other day-to-day work pressures took priority, resulting in the quality representative reacting to proposals from the researcher.

In summary, the lack of internal capability and the lack of time and resources to move the innovation forward were found to be the main obstacles. This is consistent with the key findings from the exploratory phase.

Structure capital

The researcher found that there were two practical problems within the action planning phase. First, a lack of a formalised structure and documentation system within the case study company became an obstacle in sharing information between the researcher and the ArchSME quality representative. The quality representative, for example, explained why he or she could not offer some documents which the researcher required:

> I haven't had an opportunity to dig out working copies.... I can't find QR3 or QR4.

Second, the senior management did not drive the interim project review action plan through the team structure. The initial action plan for the interim project review process innovation was solely developed by the researcher. Although the ArchSME quality representative (senior management) was involved in the development of the action plan, other task group team members from the firm did not participate.

In summary, the lack of a formalised structure and documentation system and lack of senior management driving the innovation implementation through the team structure were apparent in this phase. This is consistent with the key findings from the exploratory phase.

Relationship capital

The relationship capital in the action planning phase was located at a 'social level'. Interactions between the researcher and the ArchSME quality representative were evident in informal meetings and in telephone conversations. The importance of informal ways to carry out this innovation was emphasised. After developing the relationship with the ArchSME quality representative, the researcher found that there was no difficulty in asking for information and documentation from the company.

The researcher realised the importance of the client role for ArchSME. An introduction of a '360-degree client' perspective into an interim project review project (interim project review session) was designed to enable client interaction at both project and organisational levels. The interactions had potential to help employees to build more collaborative partnerships and understand clients' business needs in order to identify other revenue generating opportunities.

In summary, relationship capital in the action planning phase was found to be at a 'social' level and it became the main constraint to move the innovation forward. This is consistent with the key findings from the exploratory phase.

Knowledge capital

The setting up and coordination of the social knowledge capital was carried out principally within the technical knowledge capital. In a social context, knowledge capital was stimulated by face-to-face meetings and sense-making, with tacit knowledge being shared. In a technical context perspective, three mechanisms were used. First, 'the action plan' provided the clear aims, objectives and deliverables for the interim project review process innovation. Second, the use of emails helped the knowledge sharing activity between the researcher and the ArchSME quality representative prior to meeting. Also, it helped the researcher to set up meetings with the quality representative. Finally, the use of telephones in the action planning phase was important, although there were often significant delays in ArchSME staff returning calls.

In summary, knowledge capital was initially stimulated through the 'technical system' through the action plan and by communication via email and telephone. This provided the platform to commit ArchSME staff to the 'social system' meeting. This commitment, however, was limited due to higher project activity on specific projects, rather than the reallocation of resources to non-project-specific innovation. This lack of adequate and sustained commitment was a key obstacle to progress the innovation. This is consistent with the key findings from the exploratory phase.

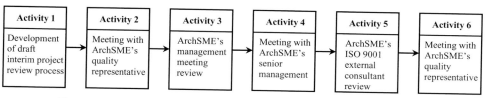

Figure 6.1   Six main activities within the action taking phase.

6.4   Action Taking

6.4.1   Practice

The action taking phase was held over a 6-month period, from the end of May 2004 to the end of November 2004. There were six main activities within this phase (see Figure 6.1). These activities are discussed in turn.

Activity 1: development of draft interim project review process

Based on the objectives of this innovation activity (see Table 6.2), the first draft of the interim project review process (including the interim project review process policy, guidelines and checklists) was developed by the researcher and sent to ArchSME's quality representative in June 2004. The interim project review process was structured into nine main sections: interim project review policy; purpose of the process; scope of the process; references; definitions, responsibility and authority; overview of the process and activity descriptions; measures; and, appendices. Each section is briefly discussed below.

The *interim project review policy* section introduced the ArchSME policy in conducting the interim project review activities, including its rationale and benefits.

The *purpose of the process* section introduced the purpose, objectives and measurement criteria of the interim project review process.

The *scope of the process* section described the scope to the interim project review process. The distinctive characteristics between 'high' focus and 'low' focus projects were made. There are three subsections under this section. The first subsection described the activities for low and high focus types of projects. The second subsection focused on illustrating the roles for low and high focus types of projects. The final subsection described the deliverables for low and high focus types of projects.

The *references* section guided the staff to the relevant ISO 9001 quality management system procedures.

The *definitions* section introduced definitions of the terms used throughout the document.

The *responsibility and authority* section described the responsibility and authority of people who participated in the interim project review process.

The *overview of the process and activity descriptions* section expressed the workflow for the interim project review process. There are six subsections under this section. Each subsection presented as an activity. The detailed work description was under each subsection.

The *measures* section was designed to give staff the measurement criteria in determining the effectiveness of interim project reviews.

The *appendices* section listed of supporting checklists for the interim project reviews. The detailed questions which made up the checklists were not developed at this time.

There were two problems in the development of the detailed checklists. First, the key challenge the researcher encountered was ensuring that the interim project review process was in line with ArchSME's ISO 9001 system. Further, checklists needed to align with the firm's work practices. This required the researcher working closely with the company's staff. Second, the researcher had to integrate two different perspectives: the first was the ArchSME senior management who were keen to have 'closed' checklists and the second was the researcher who wanted to have 'open' checklists. The rationale for the closed checklist design was that the senior management were keen to find the hard, quantitative 'indicators' to measure project performance. A closed questionnaire was, therefore, designed in response to the 'asset' view of knowledge. The initial idea to develop the interim project review process, however, was to share the project information between teams and to share tacit knowledge between people. An open questionnaire was designed to stimulate and capture soft, qualitative project performance issues. The open questionnaire was thus designed in response to the 'process' view of knowledge.

This idea led the researcher to further distinguish between 'high' and 'low' focus projects. A closed checklist was used to measure the project performance and to help management activities for both types of projects. An open question checklist included a discussion session targeted at 'high' focus projects. This approach enabled precious human capital to be targeted and leveraged at 'high focus' projects. The distinction between 'low' and 'high' focus projects is discussed in Activity 2.

The researcher decided to have a meeting with the case study firm's quality representative to move the innovation forward.

Activity 2: meetings with ArchSME's quality representative

Two meetings were held in this stage. Before the first meeting, the first draft of the interim project review process was sent to the ArchSME quality representative. The purpose of the meeting was to confirm that the focus and content of the interim project review process met the firm's requirements, access to relevant company documents, and clarify the issues raised in the previous stage. A number of issues with the first draft of the interim project review process were highlighted by the ArchSME quality representative during the meeting.

First, the criteria of the *purpose of the process* section for project performance (which were correctness, design, style, documentation and efficiency) were deleted. The quality representative gave the feedback as:

I don't think we can measure it.

Table 6.3 The distinctive characteristics between high and low focus projects.

Characteristic	High focus	Low focus
Budget	More than £ X	Less than £ X
Time to deliver	More than 1 year to operation	Less than 1 year to operation
Client involvement	No experience in the past working with this client	Good experience working with this client
Supplier involvement	No experience in the past working with this supplier	Good experience in the past working with this supplier
Team involvement	More than one project team to operation	Only one project team to operation

Second, the criteria of the *scope of the process* section for distinguishing 'low' from 'high' focus projects were adopted by the quality representative. The five criteria were budget, time to deliver, team involvement, client involvement and supplier involvement (see Table 6.3). The budget level at which a project was deemed 'high focus' was not determined.

The ArchSME quality representative agreed to the researcher's idea to distinguish low from high focus project because:

> Things like the house extension, [the project] will be smaller ... we know it's possible to run the whole [interim project review] process, ... but there is no point to do so.... We are quiet happy and easy to manage [the low focus project] ... let's just concentrate on the [high focus project].

Finally, the three subsections – activities, roles and deliverables – under the scope of the process were deleted by the quality representative. The need of the simplicity was again stressed.

Based on the results of the meeting with the quality representative, the second draft of the interim project review process was revised and renamed as the interim project review handbook by the researcher.

The second meeting took place in July 2004. Before the meeting, the second draft of the interim project review process was sent to the ArchSME quality representative. The purpose of the meeting was to confirm that the revisions made in response to the key issues raised in the first meeting met his or her requirements. Two unanswered questions/issues from the first meeting were discussed. First, the quality representative quantified the contract sums for high and low focus projects and adjusted client involvement criterion (see Table 6.4).

Second, the samples of project documents the researcher required were prepared by the ArchSME quality representative.

The ArchSME quality representative found the interim project review handbook valid and that it should be reviewed by the firm's management board. The researcher requested that he or she be present at the review meeting; however, the ArchSME quality representative refused on the basis of company policy. A revised version of the interim project review process (the third draft) was confirmed by the ArchSME quality representative through an email.

Table 6.4   The distinctive characteristics between high and low focus projects (version 1).

Changes	Characteristic	High focus	Low focus
Before	Budget	More than £ X	Less than £ X
	Time to deliver	More than 1 year to operation	Less than 1 year to operation
	Client involvement	No experience in the past working with this client	Good experience working with this client
	Supplier involvement	No experience in the past working with this supplier	Good experience in the past working with this supplier
	Team involvement	More than one project team to operation	Only one project team to operation
After	Budget	More than £50 000	Less than £50 000
	Time to deliver	More than 1 year to operation	Less than 1 year to operation
	Client involvement	Good experience working with this client (principal clients)	No experience in the past working with this client
	Supplier involvement	No experience in the past working with this supplier	Good experience in the past working with this supplier
	Team involvement	More than one project team to operation	Only one project team to operation

Activity 3: ArchSME's management meeting review

The third version of the interim project review process was reviewed at the ArchSME management meeting which took place in July 2004. The meeting had been delayed by 1 week because of work pressures within the firm. To reiterate, the action researcher was not present at this meeting, and the feedback given below is from written remarks on the tabled interim project review handbook made by all four of the firm's associate directors and one team leader. The common theme throughout the feedback was a requirement for further 'simplicity' in the interim project review process to target the practice nature of ArchSME's work. A team leader, for example, said that:

> we need to keep the processes simple to ensure take-up .... The feasibility and planning phase process should be reduced by two thirds. We just need to know who the client is, what the brief is and whether we've sent a fee letter. We also need to ensure planning conditions are signed off and that the client signs off the design. In the design and build process innovation usually occurs after tender rather than after planning. There would be no snagging meetings or certificates for making good defects.

An associate director confirmed this need for greater simplicity by commenting:

> Seems to be a very large document; lost interest by the end of page 4. Checklist looked good but too complicated – also not understood fully, so difficult to then explain to team. The checklist could prove valuable in prompting action points for other things.

As well as the concerns expressed about the complexity of the process, there was a significant debate about the alignment of the interim project review process with the work undertaken by the firm. An associate director, for example, commented that:

> Design and build and traditional contract checklists would have different questions. Post construction phase checklist could be better written in line with our business.

Similarly, a team leader said that:

> A specific innovation activity could be added to include the reclarification of the scope of works, and tracking conditions for [Building Regulations].

The key challenge the researcher encountered was securing consensus from the individuals within the meeting on how to progress the innovation. The researcher decided to arrange a follow-up meeting with senior management and the quality representative to undertake what revisions should be prioritised and to maintain senior management commitment to the interim project review process innovation.

Activity 4: meeting with ArchSME's senior management

Before the meeting with the firm's senior management, two pilot projects – one a 'high focus' type project, the other a 'low focus' type project – for testing the interim project review process were confirmed and sent by an email to the researcher by the quality representative in July 2004:

> There are two projects for which we can use . . . when do you want to hold your face-to-face meeting?

There were two meetings held in this stage. The first meeting was attended by two associate directors (senior management) and the researcher. One associate director (Participant A) was one of the respondents in the exploratory phase and participated in the company workshop. The second associate director was from team 2 (see Figure 5.2 for the structure of ArchSME). The key issues carried out of the discussion are as follows.

First, both associate directors stressed that there was a need to further simplify the interim project review process. Two issues were raised. First, the documents were too complex, as noted by the second associate director:

> We got to simplify the works. We are looking into the architectural agreement documentation which basically is a listing of who is going to do what, in what stages. So everyone is very clear about what we are going to do.

Further, the second associate director expressed that there were too many questions within the checklist. The second associate director conveyed this in the question:

> Can you make this process simple and stupid?

There was agreement between the two associate directors that the questions within the checklist were too many and too complex and needed to be reduced to two or three questions.

Second, both associate directors challenged the need to distinguish between 'low' focus and 'high' focus projects. Again, the requirement to further simplify the interim project review process was addressed. The second associate director said:

> I don't know what your thought is? For example, you asked the question like did we obtain a copy of planning permission? [This is too detailed.]

After explaining the rationale for this distinction by the researcher, both associate directors adopted the researcher's proposal.

Finally, there was a debate between the two associate directors concerning where the responsibility and authority for the interim project review process should be located. Participant A thought that the responsibility and authority for the review should be only at the architect level. The second associate director agreed that the role of the reviewer should be at the architect level, but the role of moderator and the approval authority should be at associate director level. There was no agreement between them because of time pressure – they had a meeting with clients outside of the office which they had to attend.

After the meeting with the two associate directors, the researcher had a follow-up meeting with the ArchSME quality representative. The three issues raised in the meeting with two associate directors were considered and appropriate adjustment to the interim project review process made by the researcher and the quality representative as follows.

First, the researcher and the ArchSME quality representative agreed that the need to further reduce the number of questions in the checklist from the original nine questions down to three or four questions.

Second, the researcher and the ArchSME quality representative agreed that the key indicators to distinguish between low and high focus projects needed to be driven solely by the firm's business needs. The criteria of the *scope of the process* section for distinguishing 'low' from 'high' focus projects were reduced by the ArchSME quality representative from five criteria which were budget, time to deliver, team involvement, client involvement and supplier involvement, down to one which was client involvement (see Table 6.5). The rationale was to make the process simpler. The description of 'principal clients' for high focus of client involvement criterion was deleted by the ArchSME quality representative due to sensitivity issues, i.e. accidental disclosure to clients that they were not considered as 'principal clients'.

The distinctive characteristic between the low and high focus projects for client involvement remained. The ArchSME quality representative made a comment, for example, on the criterion of supplier involvement as the process will be too complex:

> ...when you got a great, big development project, they are just so many people being involved.

Third, the responsibility and authority for different types of projects for the interim project review was made by the ArchSME quality representative. The actors and their roles in the different types of project are described in Table 6.6.

Table 6.5 The distinctive characteristics between high and low focus projects (version 2).

Changes	Characteristic	High focus	Low focus
Before	Budget	More than £50 000	Less than £50 000
	Time to deliver	More than 1 year to operation	Less than 1 year to operation
	Client involvement	Good experience working with this client (principal clients)	No experience in the past working with this client
	Supplier involvement	No experience in the past working with this supplier	Good experience in the past working with this supplier
	Team involvement	More than one project team to operation	Only one project team to operation
After	Client involvement	Good experience working with this client	No experience in the past working with this client

The meeting moved on to focus on the checklists. Each checklist was discussed. Based on the discussion, the fourth version of the interim project review process was produced and renamed as QW01 ArchSME guidelines for interim project review.

The key challenge the researcher encountered was securing consensus for, and sign-off of, the interim project review process. The researcher and the ArchSME quality representative, in line with the firm's ISO 9001 system, decided that the fourth version of the interim project review process, QW01 ArchSME guidelines for interim project review, needed to be reviewed by the firm's external ISO 9001 consultant.

Activity 5: ArchSME's external ISO 9001 consultant review

An external ISO consultancy is mainly leading ArchSME's endeavour to gain ISO 9001 accreditation. Mr. X, the company's external ISO 9001 consultant, reviewed the latest version of the interim project review handbook, and noted that the interim project review activity is the 'icing on the cake' for the customer satisfaction process and proposed the following:

(1) inclusion of an executive summary saying what the interim project review does;

Table 6.6 The responsibility and approval authority for high and low focus projects (version 1).

Roles	Types of project	Responsibility and approval authority
Moderator	High focus	Associate/team leader
	Low focus	Team leader
Reviewer	High and low focus	Job runner
Participant	High focus	Project team/other teams/directors/clients, etc.
	Low focus	Project team

(2) inclusion of operational flow charts for both 'high' and 'low' focus projects;

(3) one-to-one interviews with the client should be included in the interim project review activity; and,

(4) the feedback from the one-to-one interviews with client should be reviewed and discussed in the interim project review session.

In order to clarify and interpret these issues correctly, the researcher decided to have a meeting with the ArchSME quality representative.

Activity 6: meeting with ArchSME's quality representative

The meeting took place in July 2004. The purpose of this meeting was to discuss the changes proposed by the external ISO consultant.

The first and second suggestions were rejected by the quality representative. These two issues the external ISO consultant suggested were to be consistent with the ArchSME's ISO 9001 flow chart. However, the overview of the process and activity descriptions was already detailed in the QW01 ArchSME guidelines for interim project review.

The third suggestion, the idea of conducting one-to-one interviews with the client, was adopted by the ArchSME quality representative and it was decided to focus on 'high' focus projects. The quality representative also assigned herself to conduct the one-to-one interviews with the client work as part of his or her role as the firm's business development manager (see Table 6.7).

The final issue proposed was adopted by the ArchSME quality representative, and that the feedback from one-to-one interviews with the client would be discussed in the interim project review session. The rationale for this decision was to ensure client involvement and to further deepen the relationship with the client.

Based on these responses, the fifth version of the interim project review process was produced by the researcher and became part of the ArchSME

Table 6.7 The responsibility and approval authority for high and low focus projects (version 2).

Roles	Types of project	Responsibility and approval authority	
		After	Before (see Table 6.6)
Moderator	High focus	Associate/team leader	Associate/team leader
	Low focus	Team leader	Team leader
Reviewer	High focus	Business development/Job runner	Job runner
	Low focus	Job runner	Job runner
Participant	High focus	Project team/other teams/directors/clients, etc.	Project team/other teams/directors/clients, etc.
	Low focus	Project team	Project team

quality document system, namely the QW1 interim project review handbook (Revision A). This document was sent to the ArchSME quality representative on 21 July 2004.

The quality representative gave his or her feedback in early August 2004, which stated:

> Its mad busy here as usual and I'm conscious that I've given you no information, so rather than wait and give you a detailed explanation I am sending two [files] through and we can discuss later.

6.4.2   Reflection

Human capital

The principal variable at work in the action taking phase of innovation appeared to be the 'individual' role of the researcher. He or she realised that there were two practical problems. First, the lack of expertise and experience in developing and implementing ISO 9001 was still the major obstacle in the interim project review process innovation activity. Although the firm's ISO 9001 system has been in place from April 2004 (but not accredited), the researcher found that staff had little working knowledge and experience of the system. ArchSME's ISO 9001 system was solely developed by the external ISO consultant and that inadequate training had taken place within the case study company to build up the required ISO 9001 knowledge and capability at all levels. The researcher found that his or her role was very much the same as that of the external ISO consultant. Any good practice generated by the researcher, therefore, was not being readily absorbed by ArchSME.

Second, the lack of time by ArchSME staff to develop the interim project review process was evident in the low level of employee participation. The researcher consistently found that other task group members were extremely busy and could not find 'time' to support the innovation. The researcher found herself having to play a considerable 'championing' and 'motivating role'. The researcher had to move the process forward consistently by herself to show evidence of action and change, and in so doing, assist in envisioning and motivating ArchSME task force members.

In summary, the lack of internal capability and time to move the innovation forward were again evident. This is consistent with the key findings from the exploratory phase.

Structure capital

There were two practical problems with respect to structure capital which appeared in this phase. First, a lack of a formalised documentation system within the firm remained a constraint in sharing information between the researcher and the quality representative in real time. When the researcher asked for more samples of working documents related to the two pilot projects, for example, the quality representative said:

I am sure I can get some [the project fee letter] examples if you want.

And for at least 10 minutes, the quality representative made phone calls asking staff about the document:

I thought everyone have [these project fee letter documents].... You haven't seen them. So you don't have one of them.... Does anyone have one of copies?

By the end of the meeting, the researcher still did not receive the information.

The introduction of a formal procedure of interim project review highlighted a potential tension for small firms engaged in innovation activity. Small firms tend to have few formal processes. ISO 9000 quality management systems, however, require a significant degree of formalisation. Insistence on adherence to such formal procedures was seen to detract from the organic nature of ArchSME. In order to avoid this, the idea of 'high focus' and 'low focus' was introduced into this innovation. The proposed systems allowed for flexibility, and where possible, the interim project review process to be symbiotic with current work practices and as a consequence, 'resource light' and 'disruption free'.

Second, the lack of senior management implementation through the team structure was again evident. The researcher was the only person who mainly developed and implemented the interim project review process innovation. The ArchSME quality representative 'reactively' led the innovation activity, as her prioritises were on day-to-day, fee-income-producing projects. The ideas the researcher suggested were rarely challenged or questioned by the ArchSME quality representative.

In summary, the lack of formalised structures and documentation systems and lack of senior management in the innovation implementation activity through the team structure were found to be main obstacles in the action taking phase. This is consistent with the key findings from the exploratory phase.

Relationship capital

The relationship capital in the action taking phase was principally located in external and internal interactions at a 'social level'. Internal interactions were through informal meetings/discussions between the researcher and the quality representative, the researcher and two of the firm's associate directors. External interactions were through informal meetings between the quality representative and its external ISO consultant. In these interaction activities, the role of the researcher and external ISO consultant was to bring new knowledge and changes into the company. The role of the researcher in this project was more like that of an external consultant, rather than an embedded action researcher.

The interim project review procedure addressed the importance of the 'client involvement'. Further, this criterion in terms of the 'good experience working with this client' was defined by ArchSME as the principal distinctive characteristic between 'low' and 'high' focus projects. This stressed the importance of client interaction at an 'operational level'.

In summary, relationship capital in this phase was located at a 'social' level. The source of ideas and their application was not targeted at a specific project. The lack of operational relationship capital was found to be a key obstacle in the action taking phase. This is consistent with the key findings from the exploratory phase.

Knowledge capital

In a social context, the team-working environment tended to be in the ArchSME meeting room. The shared office environment provided the opportunities to increase interactions between the researcher, the external ISO consultant and ArchSME's senior management.

In a technical context, this took the form of emails and the telephone. First, the feedback on the interim project review process documents from other participants (such as the firm's senior management and quality representative, and its external ISO consultant) was by emails. The use of emails helped the knowledge capturing and sharing activity. Also, it helped the researcher to set up the meeting with the quality representative. Second, many of the discussions and ideas exchanged between the researcher and the quality representative were through telephone conversations.

In summary, knowledge capital was initially stimulated through the 'technical system' through the 'encoded' documents (interim project review procedure) and by communication via email and telephone. This provided the platform to commit ArchSME staff to the 'social system' meeting. The source of ideas and their application, again, did not target at a specific project. This is consistent with the key findings from the exploratory phase.

6.5   Action Evaluation

6.5.1   Practice

The QW1 interim project review handbook (Revision A) has been in place from the end of July 2004. At this time, the task force anticipated an immediate impact from the interim project review process on the effectiveness of ArchSME.

The initial external assessment for ArchSME ISO 9001 accreditation was planned for August/September 2004. By the end of July, the researcher was informed by the quality representative that the external assessment for ArchSME ISO 9001 accreditation was postponed to February 2005 due to the company workload. This argument was advanced by the ArchSME quality representative who noted that:

> Our [ISO] consultant thinks that we are not ready yet. Our system is like a new painting on the wall.

Based on the documents the quality representative sent, the sixth version of the interim project review handbook – QW1 Interim project review handbook (Revision B) – was revised by the researcher.

By the end of January 2005, the interim project review process had not been implemented.

6.5.2   Reflection

Human capital

The human capital was found to be embedded within the capacity, ability and motivation of individual (the researcher). The lack of senior management implementation, the low level of employee participation (despite have the capability of doing so) and the lack of time to develop and implement the innovation were found to be key obstacles in the interim project review process development.

The researcher believed that four principal reasons were main obstacles in the development and implementation of the interim project review process. First, the idea of the interim project review process was introduced and to certain extent, championed by the action researcher. The researcher believed that the top management did not organically and intrinsically support the interim project review process innovation. This lack of ownership of the innovative idea might well have manifested itself in the subsequent lack of senior management vision and support.

Second, senior management did not efficiently drive the interim project review process innovation into the organisation. The ArchSME's managements (senior management and middle management) were positively impressed, intrigued and motivated to pursue the proposed development approach. However, in reality, senior management did not drive the interim project review process into the organisation as it was not a prioritised project-specific, fee-earning activity. This issue led to the third issue.

Third, the lack of prioritisation the interim project review process innovation was very much a result of the lack of time allocated to the innovation. Terms like 'no time' and 'busy' were regularly mentioned.

Finally, the lack of internal capability became a constraint in the development and implementation of the interim project review process innovation. When the researcher asked for the opinion on the changes, sentences like 'I don't know', was regularly used. The discussions and meetings were principally led by the researcher. The issues and opinions the researcher suggested were rarely questioned and challenged by other participants.

In summary, the lack of top management vision, the lack of senior management support for implementation, the lack of internal capability and the lack of time variables were the main constraints in this collaborative endeavour. This is consistent with the key finings from the exploratory phase.

Structure capital

Considering structure capital, the lack of formalised structures and documentation systems and lack of senior management to drive innovation through the team structure to develop and implement the innovation activities were found to be key obstacles.

First, the lack of formalised structures and documentation systems within ArchSME was found to be the significant barrier in the interim project review process development and implementation. The lack of a formalised structure for linking and coordinating people together resulted in a loose alliance between the researcher and ArchSME. For example, information about this action research was passed within the task group members on an informal basis. In addition, the lack of documentation system to encode the issues raised in the discussions/meetings increased information/knowledge uncertainty.

Second, the lack of senior management support from inception through to implementation through the team structure at an operational level was found to be a key hurdle in the interim project review process innovation development and implementation. In summary, the lack of the formalised structure and documentation system and the lack of senior management implementation through the team structure were found to constrain the progress of the innovation. This is consistent with the key findings from the exploratory phase.

Relationship capital

The relationship capital within the interim project review process innovation development and implementation was mainly located at the 'social' level, i.e. non-project-specific innovation needs. The action research indicated that social relationship capital only helped the researcher to gain help and support to carry out her particular 'objectives' (the development of interim project review process innovation). This innovation, therefore, did not benefit from having 'operational' relationship capital to drive the innovation forward.

In summary, the lack of operational relationship capital was found to be the main obstacle in the interim project review process innovation development and implementation. This is consistent with the key findings from the exploratory phase.

Knowledge capital

The necessity for a combination of the social context and technical context was confirmed in the development and implementation of the interim project review process innovation. The knowledge capital within the interim project review process development and implementation was stimulated through the 'technical system' through the documents such as the company workshop report, the action research plan and by communication via emails, the internet and the telephone. This provided the platform to commit ArchSME staff to the 'social system' such as the company workshop and discussions/meetings in the company environment. The application of knowledge capital for the interim project review process innovation, however, did not meet a specific project need. It was more a supporting process for ArchSME ISO 9001 accreditation. This issue was considered to be the main constraint in the interim project review process innovation development and implementation.

In summary, the source of ideas and their application from a combination of social and operational contexts, which did not target at a specific project, was found to be a significant hurdle to the interim project review process innovation development and implementation. This is consistent with the key findings from the exploratory phase.

6.6   Specifying Learning

6.6.1   Practice

Specifying learning for ArchSME arguably did not happen within the action research period. The development and implementation of the interim project review activity has 'paused' at the action taking phase (see Figure 6.2).

The interim project review handbook has been in place from July 2004. The interim project review handbook was linked with ArchSME ISO 9001 external assessment schedule. The initial external assessment for ArchSME ISO 9001 accreditation was planned for August/September 2004. By the end of July 2004, the researcher was informed by the company's quality representative that the external assessment for ISO 9001 accreditation was postponed to February 2005 due to the company workload. By the end of January 2006, the interim project review handbook has not been implemented.

At this moment, ArchSME have not captured any learning from the implementation of the interim project review process as it has not been operated in a real world project setting.

6.6.2   Reflection

The following summarises the key reflections of the action research process. The purpose of specifying learning, as shown in Figure 6.3, is to draw

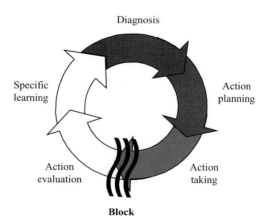

Figure 6.2   Learning block for ArchSME.

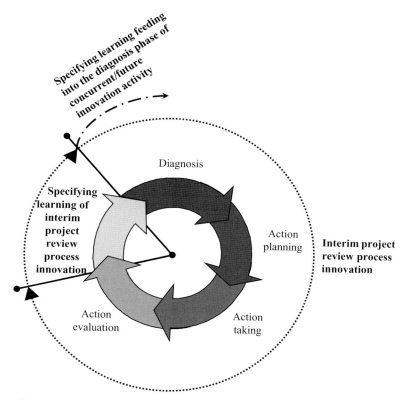

Figure 6.3   The specifying learning for the researcher.

generic lessons which can feed into subsequent (or concurrent) innovation activity.

There are two generic specifying learning themes for the researcher. First, learning blocks within ArchSME with regard to the development and implementation of the interim project review process are given (see Section 6.5.2). Second, mechanisms to overcome these blocks for concurrent/future innovation activity are offered.

Human capital

Four key human capital variables emerged from the interim project review process innovation: lack of top management 'championing', senior management not driving the implementation through the team structure, low level of employee buy in and lack of time to develop and implement the innovation activity (see Section 6.5.2). The specifying learning for concurrent/future innovation is that these four key human capital variables should be appropriately addressed to bring about successful innovation activity. They are discussed below.

First, top management did not organically and intrinsically champion and support the interim project review process innovation. The interim project review process idea did not directly come from senior management vision; rather, it came principally from the researcher. This lack of ownership of the innovative idea might well have manifested itself in the subsequent lack of 'championing' of the innovation.

Second, senior management did not drive the implementation through the team structure, which resulted in the low level of employee participation. The specifying learning for further innovation activity is that senior management must drive, and seen to be driving, the innovation from inception through to implementation. The senior management commitment and involvement also would encourage staff to get involved in the innovation activity.

Third, inadequate resources were dedicated to the innovation because of full resource allocation to day-to-day fee-income-producing project activity. Innovation activity needs to be appropriately promoted and resourced, without this innovation will whither, and staff will increasingly view non-project-specific future innovation activity as doomed to failure.

Finally, the company lacked the necessary internal capability in ISO 9000 quality management systems to locate and develop the interim project review process innovation. Innovation activity needs to have adequate capability; if this is not present in the firm, the necessary capability needs to be recruited or developed internally through training and development; or, relevant external expertise brought in. In the case of external expertise, effort should be made to transfer this capability to firm staff so that it is available after the external agent has gone.

In summary, top management championing and support, senior management implementation, the allocation of resources and the ownership of innovation are the main key variables to progressing innovation activity.

Structure capital

The key structure capital variable identified from the interim project review process innovation was the lack of the formalised structures and documentation systems (see Section 6.5.2). The specifying learning for concurrent/future innovation is that this key structure capital variable must be adequately addressed for successful innovation activity.

There is a need for adequate formalised structures and documentation systems to develop and implement innovation activity. First, a formalised structure enables roles and responsibility to be clearly assigned to progress the innovation. This formalisation legitimises the innovation through positional power or authority to capture the rationale and necessary information for the innovation and to share that information is required. Second, formalised documentation systems in place with, an inevitable result, good practice and lessons learned will be captured and shared for future use. Further, the formalisation must be balanced with a need to keep any process 'resource light', and to be sympathetic to current work practices.

Relationship capital

The lack of operational relationship capital was identified as key variable from the interim project review process innovation (see Section 6.5.2). The specifying learning for concurrent/future innovation is that this key relationship capital variable must be present for successful innovation activity. The specifying learning for further innovation is that the innovation activity has to be tangibly linked to project activity. The operational relationship capital (i.e. project-specific needs) allows the project work to be organised and controlled by appropriate individuals with responsibility.

Knowledge capital

A combination of social and technical knowledge capital channelled to a specific project was identified as key variable from the interim project review process innovation from a knowledge capital perspective (see Section 6.5.2). Innovation supported by technical knowledge capital inadequately generates, shares, leverages and exploits tacit knowledge possessed by knowledge workers.

6.7   Summary and Link

This chapter has presented the key findings from the action research phase of the case study. The next chapter brings together the key results from the exploratory phase and the action research phase of a case study to test the hypotheses set out in Chapter 3.

7

Discussion

7.1 Introduction

This chapter presents a discussion of key results from the exploratory phase (see Chapter 5) and the action research phase (see Chapter 6) of the case study. The knowledge-based innovation concept model (see Figure 3.1) proved to be useful in both understanding innovation (the exploratory phase of case study) and managing innovation activity (the action research phase of case study). Two principal types of innovation were identified in the exploratory phase of the case study: explorative innovation (see Section 5.5) and exploitative innovation (see Section 5.6). An exploitative innovation – interim project review process innovation – was the focus of the action research phase of the case study (see Chapter 6).

7.2 Types of Knowledge-Based Innovation

Two types of innovation within the company were identified as explorative innovation (see Section 5.5) and exploitative innovation (see Section 5.6). The two types of innovation were found to be a useful and valid way of understanding knowledge-based innovation. The research findings indicate that firms achieve short-term 'project-based' success with explorative innovation (see Figure 7.1, mode 1) and potential long-term 'organisational' success with exploitative innovation (see Figure 7.1, mode 2).

Explorative innovation (mode 1) focuses on client facing, specific project needs (external fee-income-producing project), resulting in effective and efficient delivery of services to satisfy current external project needs; whilst exploitative innovation (mode 2) focuses on organisational and general client development activity, resulting in organisational effectiveness and efficiency improvement, and in so doing, potentially generating sustainable competitive advantage. The distinctive feature of exploitative innovation (compared to explorative innovation) is that new phenomena, systems or structures are more readily embedded in the structure capital of the firm. In contrast, explorative innovation tends to rotate around specific projects and the lessons

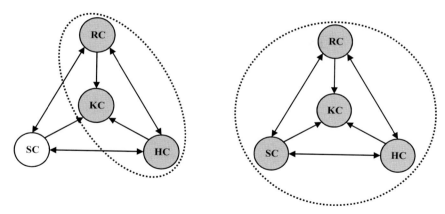

Mode 1: Explorative innovation

Mode 2: Exploitative innovation

Figure 7.1   Types of knowledge-based innovation.

learned are often not effectively encoded into the structure capital of the firm for subsequent retrieval and use.

The next section tests the research hypotheses set out in Chapter 3 on the basis on the data gathered and analysed in the case study (see Chapters 5 and 6).

7.3   Hypothesis 1: Knowledge-Based Resources

The first hypothesis posed in Section 3.4 was concerned with knowledge-based resources.

Hypothesis 1

A small construction professional practice which develops integrated individual, organisational and client human capital, structure capital and relationship capital will generate a more appropriate stock of resources for successful innovation.

Hypothesis 1 consists of three sub-hypotheses. They are discussed below. At the end of this section, Hypothesis 1 will be discussed (see Section 7.3.4).

7.3.1   Hypothesis 1-1: human capital

Hypothesis 1-1

A small construction professional practice which develops integrated individual, organisational and client human capital will generate a more appropriate stock of human capital resources which will contribute to successful innovation.

The results from the exploratory phase and the action research phase provide general support for Hypothesis 1-1.

Explorative innovation

The human capital for explorative innovation identified in the exploratory phase was embedded within the capacity, ability and motivation of staff (see Section 5.5.1) and external supply chain partners (see Section 5.5.3).

In successful explorative innovation, human capital was focused on a specific project at an 'operational' level with knowledge being elicited, mobilised and integrated from individual, organisational and client 'social' and 'operational' sources to progress project challenges with innovative solutions, e.g. new designs (innovation 3) (see Section 5.5.1). It was found that successful explorative innovation mainly relied on staff (including individual knowledge workers, management and the client) working together through the team structure. The tangible and immediate project focus gave the innovation activity sufficient priority to secure adequate commitment and resources to ensure its success.

In contrast, in unsuccessful explorative innovation, knowledge from individual, organisational and client human capital tended to be located at a non-project-specific 'social' level, rather than at a project-specific 'operational' level, e.g. new materials (innovation 6) (see Section 5.5.1). These bodies of knowledge, without an integrating project context, were characterised as being disjointed with each other. Innovation activity from these sources, unless brought together and reconfigured to meet the needs of a particular project, lacked the prioritisation and legitimisation to claim resources to bring about successful innovation.

In summary, explorative innovation Hypothesis 1-1 was confirmed, i.e. successful innovation was characterised by integrated, operational human capital around a focal project context; whilst unsuccessful innovation was evidenced by disjointed bodies of knowledge located at a social level without the benefit of a focal project to stimulate purposeful action.

Exploitative innovation

The human capital for exploitative innovation in the exploratory phase was located within the capacity, ability and motivation of staff (see Section 5.6.1) and external supply chain partners (see Section 5.6.3), particularly clients and suppliers.

Successful exploitative innovation was found to have the benefit of senior management who were sufficiently motivated to drive the innovation through the team structure to successful implementation, and to encourage appropriate employee participation in the process (see Section 5.6.1). First, the senior management role was seen as very much encouraging the integration of individual and organisational human capital through appropriate project teamwork. Projects for exploitative innovation were reviewed as 'internal' projects (rather than 'external' fee-producing projects). Once this teamwork was in place, individual knowledge workers engaged with client human capital within the context of a specific project, e.g. mission statement (innovation 1), Investors in People (IiP) (innovation 2) and company restructure (innovation 4) (see Section 5.6.1). The key factor in successful exploitative innovation was senior management involvement in 'implementation activity'. The research findings indicate that in idea creation, senior management has a boundary-spanning role as they have sufficient knowledge of everyone's work, the firm, customers, suppliers and the industry to be able to integrate the diverse views of the stakeholders, and to come up with appropriate ideas. There were four important dimensions to the role of senior management in driving and implementing innovation activities: the allocation of project work into the team, teamwork supervision, the training and development of staff and the motivation of staff to participate in innovation activity.

In contrast, unsuccessful exploitative innovation was characterised by four key human capital variables: lack of top management 'championing' of the innovation, senior management not driving the implementation of the innovation through the team structure, low level of employee participation and lack of time for staff to develop and implement the innovation activity, e.g. seminars (innovation 5) and Learndirect project (innovation 7) (see Section 5.6.1). The key findings were confirmed in the action research phase. They are discussed below.

First, the key role of senior management in framing and prioritising innovation activity was confirmed as a key human capital variable for exploitative innovation success (see Section 6.2.2). This provided the innovation activity with the necessary 'championing' to forge and resource the bringing together of individual and organisational human capital (see Sections 6.5.2 and 6.6.2). When the new idea did not directly come from the senior management, the motivation to champion the innovation was seen to be weaker (see Section 6.6.2).

Second, senior management did not drive the implementation through the team structure, which resulted in the third factor, low levels of employee participation (see Sections 6.5.2 and 6.6.2). Senior management must drive, and seen to be driving, the innovation from inception through to implementation. The senior management commitment and involvement was seen to encourage staff to get involved in the innovation activity.

Third, the lack of the internal capability was confirmed to be a key constraint to progress innovation (see Sections 6.3.2, 6.4.2, 6.5.2 and 6.6.2). Innovation activity needs to have adequate capability; if this is not present in the firm, the necessary capability needs to be recruited in, or developed internally through training and development or relevant external expertise brought in.

Finally, the lack of the capacity to ensure adequate allocation of time and resources to move the innovation forward was viewed as a key constraint to progress innovation (see Sections 6.3.2, 6.4.2, 6.5.2 and 6.6.2). Innovation activity needs to be appropriately promoted and resourced. Otherwise, it was observed that company resources were allocated solely to day-to-day fee-income-producing project activity.

In summary, exploitative innovation Hypothesis 1-1 was confirmed, i.e. successful innovation was characterised by integrated, operational human capital around a tangible, client-driven business need; whilst unsuccessful innovation was evidenced by disjointed bodies of knowledge located at a social level without the benefit of an integrating client-driven business need.

Summary

In combination, the findings for explorative and exploitative innovation support Hypothesis 1-1, and indicate the positions shown in Figure 7.2. The left-hand side of diagram depicts successful innovation supported by an integrated, dynamic 'operational' project and/or client-driven business human capital locus. In contrast, the right-hand side of diagram indicates that where there is no specific project or client-driven business focus, innovation fails because of disjointed and unfocused bodies of knowledge residing in individual, organisational and client human capitals at a social level.

7.3.2   Hypothesis 1-2: structure capital

> **Hypothesis 1-2**
>
> A small construction professional practice which develops integrated individual, organisational and client structure capital will generate a more appropriate stock of structure capital resources which will contribute to successful innovation.

The analysis of the data for explorative innovation from the exploratory phase does not provide evidence to support for Hypothesis 1-2. On the other hand, the data for exploitative innovation from the exploratory phase and the action research phase provide broad support for Hypothesis 1-2.

Explorative innovation

The structure capital for explorative innovation identified in the exploratory phase was the creation and maintenance of appropriate team structures to enable purposeful and productive project-based teamwork. There was no 'quantitative' innovation performance measurement system to determine the success of innovation activity (see Section 5.5.2).

In successful explorative innovation, structure capital was found to have enduring senior management support for the setting up and maintenance of enabling team structures from inception through to implementation which stimulated and developed team-based ideas at an operational level (see Section 5.5.2). Two issues were raised. First, the senior management was seen to be the key enabler to bring together individual structure capital through the organisational team structure and to promote their engagement with client structure capital within the context of a specific project. Second, the team and communication structures encouraged and enabled ideas to be generated,

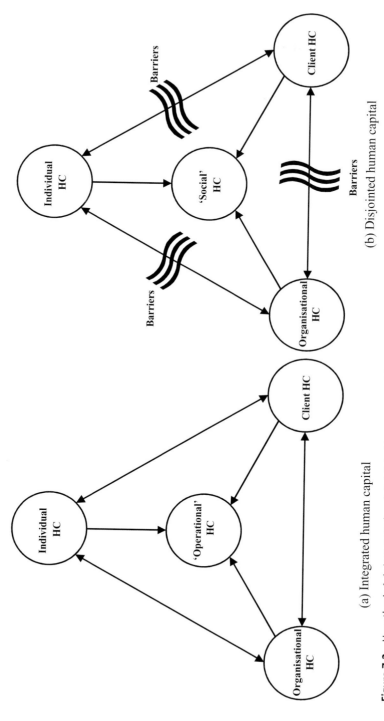

(a) Integrated human capital

(b) Disjointed human capital

Figure 7.2   Hypothesis 1-1: Integrated and disjointed human capital for explorative and exploitative innovation.

progressed and integrated from individual and external supplier chain partners' structure capital to create team-based ideas to feed into specific project needs. It was found that the new knowledge and practices flowing from explorative innovation are often not embedded in the organisational structure capital due to management attention and company resources being constantly focused on current or future project-specific needs.

In contrast, in unsuccessful explorative innovation, individually created ideas, derived from his or her 'social' relationship capital, were found to be inappropriate for specific project needs, and were pursued relatively independently of the team structure, e.g. new materials (innovation 6) (see Section 5.5.2). Three issues were raised. First, without an integrating project hub, senior management did not commit to setting up and maintaining appropriate structures to support the innovation activity. Second, without these team and communication structures, individual structure capital was separate from organisational and client structure capital. The individual, organisation or external supplier chain partners' structure capital did not become embedded at an operational structure capital level. Finally, the lack of specific structure capital was seen to limit the amount of relevant information within the organisational structure capital (see Section 5.5.2).

In summary, for explorative innovation Hypothesis 1-2 was falsified. Evidence shows that there was no integrated individual, organisational and external supply chain partners' structure capital within successful explorative innovation.

Exploitative innovation

The structure capital for exploitative innovation within the exploratory phase was embedded within formalised administrative systems, team structures and computer systems. There were no quantitative innovation performance measurement systems (see Section 5.6.2).

The successful exploitative innovation was found to have formalised structures and documentation systems, enduring senior management support from inception through to implementation and supported by an enabling team structure which stimulated and developed team work at an operational level, e.g. mission statement (innovation 1), IiP (innovation 2) and company restructure (innovation 4) (see Section 5.6.2). In all these innovations, the principal focus was to develop the structure capital in some way. The success of exploitative innovation was seen to be dependent on formalised structures and documentation systems. In addition, senior management support through 'the team structure' was essential for driving and implementing innovation activities such as the allocation of project work into the team and the supervision of teamwork.

In contrast, the unsuccessful exploitative innovation was found to have no formalised structures and documentation systems and no senior management support to drive the innovation down into the organisation, e.g. seminars (innovation 5) and the Learndirect project (innovation 7) (see Section 5.6.2). The key findings were confirmed in the action research phase. They are discussed below.

First, the key role of the formalised structures and documentation systems and the key role of senior management endeavour in driving the innovation implementation through the organisational team structure were confirmed as the key factors to develop and implement innovation activity (see Section 6.2.2). Further, it was emphasised that formalisation must be balanced with a need to keep any process 'resource light', and to be sympathetic to current organisational structure capital (see Section 6.6.2).

Second, the lack of a formalised structure and documentation system was confirmed to be a key constraint to progress innovation (see Sections 6.3.2, 6.4.2, 6.5.2 and 6.6.2). The need for a formalised structure to enable roles and responsibilities to be clearly assigned to progress the innovation, and the need for formalised documentation systems to capture and share good practice and lessons learned for future use, was confirmed as a critical element for the success of the interim project review process innovation (see Section 6.6.2).

Finally, the lack of senior management in the innovation implementation activity through the organisational team structure was found to be an obstacle in the progression of the innovation activity (see Sections 6.3.2, 6.4.2, 6.5.2 and 6.6.2). Senior management was seen as having a key role in bringing together individual, organisational and external supplier chain partners' structure capital to progress specific project needs.

In summary, exploitative innovation Hypothesis 1-2 was confirmed, i.e. successful innovation was characterised by integrated, operational structure capital around a tangible, client-driven business need; whilst unsuccessful innovation was evidenced by disjointed structures and encoded knowledge located at a social level without the benefit of an integrating client-driven business need.

Summary

Figure 7.3a shows successful exploitative innovation supported by an integrated, dynamic 'operational' project and/or client-driven business structure capital locus. In contrast, Figure 7.3b presents where there is no specific project or client-driven business focus, innovation fails because of disjointed and unfocused structures and encoded knowledge residing in individual, organisational and external supply chain partners' structure capitals at a social level.

7.3.3   Hypothesis 1-3: relationship capital

Hypothesis 1-3

A small construction professional practice which develops integrated individual, organisational and client relationship capital will generate a more appropriate stock of relationship capital resources which will contribute to successful innovation.

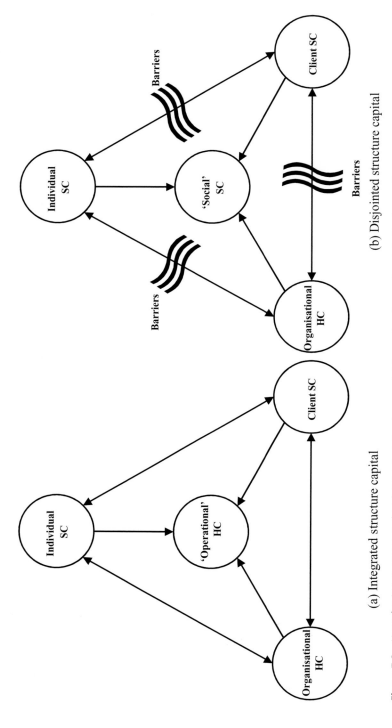

(a) Integrated structure capital

(b) Disjointed structure capital

Figure 7.3 Hypothesis 1-2: Integrated and disjointed structure capital for explorative and exploitative innovation.

The analysis of the data from the exploratory phase and the action research phase provides general support for Hypothesis 1-3.

Explorative innovation

The relationship capital for explorative innovation within the exploratory phase was located at internal and external supply chain partners' interaction domains of activity, particularly clients and suppliers (see Section 5.5.3).

In successful explorative innovation, knowledge from individual, organisational and client 'operational' and 'social' relationship capital sources was integrated and fed into specific project needs, e.g. new designs (innovation 3) (see Section 5.5.3). It was found that rich resources of relationship capital provided the variety of new ideas to fuel successful explorative innovation. For example, within a project context, knowledge from external supplier chain partners (e.g. suppliers) was used in specific projects.

In contrast, unsuccessful explorative innovation was underpinned solely by 'social' relationship capital sources which did not meet project-specific innovation needs – be they 'external' fee-income producing projects or 'internal' project to promote organisational and general client development activity, e.g. new materials (innovation 6) (see Section 5.5.3). The bodies of knowledge from individual, organisational and external supply chain partners' relationship capital, without an integrating project context, were characterised as being disconnected with each other.

In summary, explorative innovation Hypothesis 1-3 was confirmed, i.e. successful innovation was characterised by integrated, operational relationship capital around a focal project context; whilst unsuccessful innovation was evidenced by disjointed bodies of relationship knowledge located at a social level without the benefit of an integrating project focus.

Exploitative innovation

The relationship capital for exploitative innovation within the exploratory phase was located at internal and external supply chain partners' interaction domains of activity, particularly clients, suppliers and business advisers (see Section 5.6.3). The role of the 'business adviser' was particularly stressed in exploitative innovation. The business adviser seems to be an important source of knowledge and information external to the company. The need for the company to be appropriated involved in such external business networks is thus especially important, as it often does not have the knowledge and resource needed to develop innovations on their own. The business advisers advised on generic company strategy and organisation rather than specific architectural practice issues.

In successful exploitative innovation, knowledge from individual, organisational and client 'operational' and 'social' relationship capital sources was integrated and fed into specific project needs, e.g. mission statement (innovation 1), IiP (innovation 2) and company restructure (innovation 4) (see Section 5.6.3). In contrast, unsuccessful exploitative innovation was underpinned

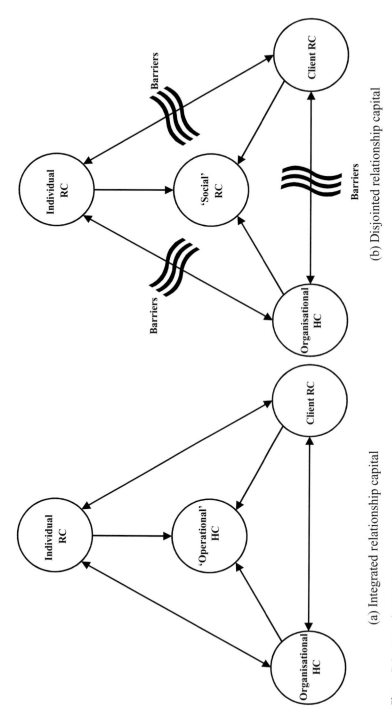

(a) Integrated relationship capital

(b) Disjointed relationship capital

Figure 7.4   Hypothesis 1-3: Integrated and disjointed relationship capital for explorative and exploitative innovation.

solely by 'social' relationship capital sources which did not feed into project-specific innovation, rather they fed into non-project-specific innovation (such as organisational and general client development activity), e.g. seminars (innovation 5) and the Learndirect project (innovation 7) (see Section 5.6.3). The key findings were confirmed in the action research phase.

The relationship capital for exploitative innovation within the action research phase was located at 'social' level, i.e. non-project-specific innovation needs (see Sections 6.2.2, 6.3.2, 6.4.2, 6.5.2 and 6.6.2). The lack of operational relationship capital was confirmed as the key obstacle for the success of the interim project review process innovation (see Sections 6.2.2, 6.3.2, 6.4.2, 6.5.2 and 6.6.2). The innovation activity has to be tangibly linked to project activity (i.e. project-specific needs) which brought together individual, organisational and client operational relationship capital (see Section 6.6.2).

In summary, exploitative innovation Hypothesis 1-3 was confirmed, i.e. successful innovation was characterised by integrated, operational relationship capital around a tangible, client-driven business need; whilst unsuccessful innovation was evidenced by disjointed bodies of relationship knowledge located at a social level without the benefit of an integrating client-driven business need.

Summary

In combination, the findings for explorative and exploitative innovation support Hypothesis 1-3, and indicate the positions shown in Figure 7.4. The left-hand side of diagram depicts successful innovation supported by an integrated, dynamic 'operational' project and/or client-driven business relationship capital locus. In contrast, the right-hand side of diagram indicates where there is no specific project or client-driven business focus, innovation fails because of disjointed and unfocused bodies of knowledge residing in individual, organisational and client relationship capitals at a social level.

7.3.4   Comment on Hypothesis 1

Hypothesis 1 examines the knowledge-based resources for innovation. Table 7.1 summarises the outcome of the testing of the hypothesis.

Table 7.1 Summary of Hypothesis 1.

	Testing results (confirmed/falsified)	
Hypothesis	Explorative innovation	Exploitative innovation
H 1: knowledge-based resources	Falsified	Confirmed
H 1-1: HC	Confirmed	Confirmed
H 1-2: SC	Falsified	Confirmed
H 1-3: RC	Confirmed	Confirmed

HC, human capital; SC, structure capital; RC, relationship capital.

The research findings present a varied picture depending on whether the innovation was explorative or exploitative in nature. For exploitative innovation, Hypothesis 1 was confirmed. Successful exploitative innovation was characterised by integrated individual, organisational and client human capital (see Section 7.3.1), structure capital (see Section 7.3.2) and relationship capital (see Section 7.3.3) around a specific client-driven need. The unsuccessful exploitative innovation was characterised by fragmented and unfocused individual, organisational and client human capital (see Section 7.3.1), structure capital (see Section 7.3.2) and relationship capital (see Section 7.3.3), and did not benefit from a specific client-driven need.

For explorative innovation, Hypothesis 1 appeared to be falsified. Successful explorative innovation is characterised by integrated individual, organisational and client human capital (see Section 7.3.1) and relationship capital (see Section 7.3.3) around a specific project. The need for integrated individual, organisational and client structure capital (see Section 7.3.2) was found not to be a prerequisite for successful innovation.

The next section describes Hypothesis 2 related to capabilities.

7.4   Hypothesis 2: Capabilities

The second hypothesis posed in Section 3.4 was concerned with capabilities.

Hypothesis 2

A small construction professional practice which generates and integrates exploitative and explorative capabilities through appropriate interaction between human capital, structure capital and relationship capital will generate appropriate knowledge capital to stimulate and support successful innovation.

Hypothesis 2 consists of three sub-hypotheses. They are discussed below. At the end of this section, Hypothesis 2 is discussed (see Section 7.4.4).

7.4.1   Hypothesis 2-1: link between human capital and relationship capital

---

Hypothesis 2-1

A small construction professional practice which generates and integrates exploitative and explorative capabilities through appropriate interaction between relationship capital and human capital will make a positive contribution to knowledge capital.

The analysis of the data from the exploratory phase and the action research phase provides broad support for Hypothesis 2-1.

Explorative innovation

In the exploratory phase of the case study, successful explorative innovation was supported by explorative capability generated by relationship capital and human capital interaction at an 'operational' level, e.g. new designs (innovation 3) (see Sections 5.5.1 and 5.5.3). It was evident that knowledge workers actively drew upon their operational and social relationship capital sources to acquire information and knowledge that was relevant to current specific projects.

In unsuccessful explorative innovation, it was demonstrated that there was inadequate operational explorative capability generated by relationship capital and human capital; rather, the interaction was at a social level decoupled from the specific needs of a project, e.g. new materials (innovation 6) (see Sections 5.5.1 and 5.5.3).

In summary, explorative innovation Hypothesis 2-1 was confirmed, i.e. successful innovation was supported by explorative capability generated by operational relationship capital and human capital interaction around a specific project context; whilst unsuccessful innovation was evidenced by disjointed interaction between social relationship capital and human capital in non-specific project domains.

Exploitative innovation

In the exploratory phase of the case study, in successful exploitative innovation, exploitative capability was demonstrated when relationship capital and human capital was engaged with operational project, client-driven business needs, e.g. mission statement (innovation 1), IiP (innovation 2) and company restructure (innovation 4) (see Sections 5.6.1 and 5.6.3). It was found that ideas for successful exploitative innovation came from 'operational' and 'social' relationship capital sources which were fed into a specific project (innovation 1 and innovation 3 were used to support innovation 2: IiP accreditation project).

In unsuccessful exploitative innovation, there was inadequate exploitative capability generated by relationship capital and human capital interaction at an operational level; rather, it tended to be located at a 'sterile' social level which was viewed by staff as not being relevant for their immediate project work, e.g. seminars (innovation 5) and the Learndirect project (innovation 7) (see Sections 5.6.1 and 5.6.3). The key findings were confirmed in the action research phase. They are discussed below.

The lack of exploitative capability brought about by inadequate and inappropriate relationship capital and human capital interaction at an operational level within the action research phase was confirmed as a key obstacle for the success of the interim project review process innovation (see Sections 6.2.2, 6.3.2, 6.4.2, 6.5.2 and 6.6.2). Two mechanisms were identified as core constraints for this innovation success. First, the idea of the interim project review process did not come from 'operational' relationship capital source; rather, it came principally from the researcher, i.e. from an external 'social' relationship capital source. This lack of ownership by the senior management of the innovative idea manifested itself in the subsequent lack of ArchSME senior management 'championing' of the innovation. Second, there was a lack of appropriate internal human capital capability in quality management systems which resulted in the company having to rely on buying in relevant external expertise. It was found that there was little motivation to set up appropriate mechanisms to successfully transfer and develop this capability into the firm's internal human capital. The absence of appropriate knowledge transfer and internal human capital generation with respect to quality management systems exposes the firm to not having sufficient internal capability to operate, maintain and further develop its quality management systems once the external sources of capability are not present (in this case, the external quality consultant and the researcher).

In summary, exploitative innovation Hypothesis 2-1 was confirmed, e.g. successful innovation was supported by exploitative capability generated by operational relationship capital and human capital interaction around a tangible, client-driven business need; whilst unsuccessful innovation was supported by exploitative capability generated by social relationship capital and human capital interaction around an intangible, non-client-driven business need.

Summary

In combination, the findings for explorative and exploitative innovation confirm Hypothesis 2-1. The key finding indicates that successful innovation supported by operational explorative and/or exploitative capabilities are targeted at, and stimulated by, project and/or client-driven business needs; whilst in unsuccessful innovation, interaction between human capital and relationship capital was located at a social level, rather than at an operational level.

7.4.2   Hypothesis 2-2: link between structure capital and human capital

> **Hypothesis 2-2**
>
> A small construction professional practice which generates and integrates exploitative and explorative capabilities through appropriate interaction between structure capital and human capital will make a positive contribution to knowledge capital.

The analysis of the data for explorative innovation from the exploratory phase does not provide evidence to support for Hypothesis 2-2. In contrast, the data for exploitative innovation from the exploratory phase and the action research phase provide wide support for Hypothesis 2-2.

Explorative innovation

In the exploratory phase of the case study, there was no clear evidence that successful explorative innovation was supported by explorative capability generated by structure capital and human capital interaction, e.g. new designs (innovation 3) (see Sections 5.5.1 and 5.5.2). The knowledge gleaned from operational and social relationship capital sources was appropriately filtered and configured to meet a specific project need through the team structure. The senior management supported, through the team structure, the innovation activity from the idea creation to its implementation. This structure capital for explorative innovation, however, was fragile and temporary and not embedded within the company, i.e. the interaction between human capital and structure capital ended when the project finished. This does not imply that case firm's staff did not use explicit, codified material in creating knowledge; indeed, they frequently developed notes, drawings, designs and so forth. However, this material was used for the specific project only, but was not encoded, or tacit knowledge transfer mechanism enabled, within the organisational structure that allowed this knowledge to be reused by the originating team or the other three project teams within the company.

Similarly, there was evidence that unsuccessful explorative innovation was characterised by inappropriate explorative capability generated by structure capital and human capital interaction, e.g. new materials (innovation 6) (see Sections 5.5.1 and 5.5.2). The knowledge for explorative innovation came from individuals from his or her 'social' relationship capital which was not adequately transformed to meet the need of a specific project, and which was pursued relatively independently of the team. The knowledge, therefore, was not embedded into the organisational structure.

In summary, explorative innovation Hypothesis 2-2 was falsified. There was no clear evidence that successful explorative innovation was supported by explorative capability generated by human capital and structure capital interaction.

Exploitative innovation

In the exploratory phase of the case study, in successful exploitative innovation, exploitative capability was evident when human capital and structure capital were engaged with internal project, client-driven business needs, e.g. mission statement (innovation 1), IiP (innovation 2) and company restructure (innovation 4) (see Sections 5.6.1 and 5.6.2). A formalised structure and documentation system was perceived to the useful ways of capturing information and knowledge in published material such as company quality manual or company handbook. These materials were integrated by knowledge workers to acquire knowledge and information at an operational level. It was evident that when the case study firm documented knowledge in a systematic way, staff were more aware of the knowledge and could readily access it. Through the formalised structure, staff were able to share their knowledge and expertise.

In unsuccessful exploitative innovation, exploitative capability was generated by structure capital and human capital interaction at a social level, e.g. seminars (innovation 5) and the Learndirect project (innovation 7) (see Sections 5.6.1 and 5.6.2). Without a project focus, these exploitative innovations failed. The key findings were confirmed in the action research phase.

The lack of exploitative capability brought about by inadequate structure capital and human capital interaction at an operational level identified in the action research phase was confirmed as a key obstacle for the interim project review process innovation success (see Sections 6.2.2, 6.3.2, 6.4.2, 6.5.2 and 6.6.2). The interaction between human capital and structure capital was focused at a social level (see Section 6.6.2). Two core elements were identified. First, the lack of senior management commitment and involvement in the innovation activity through the team structure resulted in low levels of employee participation. The knowledge worker prioritised his or her efforts into day-to-day fee-income producing project activity, rather than engaging with the internal organisation development activity. (Indeed, this prioritisation of project work over general 'organisational development' was reinforced by individual performance being assessed against project delivery criteria – see Sections 6.4.2 and 6.5.2.) Second, the lack of internal capability (human capital) in the firm, the development of innovation activity (such as ISO 9001 quality management system or the interim project review procedure) was mainly carried out by external supplier chain partners (i.e. the researcher and ArchSME's external ISO consultant). It was found that there was no appropriate mechanisms (e.g. training) set-up to transfer this capability into the firm's internal human capital. The consequence of this is that the firm has difficulty in operating, maintaining and further developing its quality management systems once the external business advisers were not present.

In summary, exploitative innovation Hypothesis 2-2 was confirmed, e.g. successful innovation was supported by exploitative capability generated by operational structure capital and human capital interaction around a tangible, client-driven business need; whilst unsuccessful innovation was evidenced by disjointed interaction of social structure capital and human capital around a non-client-driven business need.

7.4.3   Hypothesis 2-3: link between relationship capital and structure capital

> **Hypothesis 2-3**
>
> A small construction professional practice which generates and integrates exploitative and explorative capabilities through appropriate interaction between relationship capital and structure capital will make a positive contribution to knowledge capital.

The analysis of the data for explorative innovation from the exploratory phase does not provide evidence to support for Hypothesis 2-3. In contrast, the data for exploitative innovation from the exploratory phase and the action research phase provide broad support for Hypothesis 2-3.

Explorative innovation

In the exploratory phase of the case study, there was no clear evidence that successful explorative innovation was supported by explorative capability generated by relationship capital and structure capital interaction at an operational level, e.g. new designs (innovation 3) (see Sections 5.5.2 and 5.5.3). The knowledge from 'operational' and 'social' relationship capital sources was channelled into specific project needs. This knowledge was mobilised to produce innovation within a specific project context, but was not tangibly embedded within the structural capital of the firm for future retrieval and use. Notwithstanding this lack of linkage, the innovation within the context of the project was deemed successful. Any lessons learned from project-based innovation were very much located within individual workers. Knowledge transfer between individuals at a socialisation level (see Section 2.5.3) to develop knowledge capital, to a more limited extent, was evident within individual teams. However, the fairly rigid team structure within ArchSME, where teams consisted of stable, fixed membership (see Section 5.2), created a significant barrier to informal knowledge transfer between teams. The seminar (innovation 5) was an attempt to provide a mechanism to encourage such transfer, but the lack of specific project focus led to this innovation being unsuccessful.

Similarly, there was no clear evidence that unsuccessful explorative innovation was characterised by explorative capability generated by structure capital and relationship capital interaction, e.g. new materials (innovation 6) (see Sections 5.5.2 and 5.5.3). The unsuccessful explorative innovation was underpinned solely by 'social' relationship capital sources which did not meet project-specific innovation needs. The relationship capital was presented as enabling conditions for knowledge creation and sharing. External supplier chain partners (e.g. suppliers) (see Section 5.5.3) were found to be an important source of new ideas. Without a project context, the knowledge sharing and creation only happened when a member of staff asked for advice. It was found that the knowledge worker within the firm was learning internally from colleagues.

In summary, explorative innovation Hypothesis 2-3 was falsified. There was no clear evidence that there is a link between relationship capital and formal structure capital. Successful explorative innovation was not dependent on strong human capital and formal structure capital interaction.

Exploitative innovation

In the exploratory phase of the case study, exploitative capability for successful exploitative innovation was evident when relationship capital and structure capital were engaged with internal project, client-driven business needs, e.g. mission statement (innovation 1), IiP (innovation 2) and company restructure (innovation 4) (see Sections 5.6.2 and 5.6.3). The successful exploitative innovation activity was tangibly linked to a specific project activity. The operational relationship capital allows the project work to be organised and controlled by appropriate individuals with responsibility through the organisation structure.

In contrast, in unsuccessful exploitative innovation, it was evident that there was inappropriate exploitative capability generated by relationship capital and structure capital interaction at a social level, e.g. seminars (innovation 5) and the Learndirect project (innovation 7) (see Sections 5.6.2 and 5.6.3). The key findings were confirmed in the action research phase.

The lack of exploitative capability brought about by inadequate and inappropriate structure capital and relationship capital interaction at an operational level was confirmed as the critical constraint for the interim project review process innovation success (see Sections 6.2.2, 6.3.2, 6.4.2, 6.5.2 and 6.6.2).

In the action research phase, the interaction between relationship capital and structure capital was located at a social level. This tended to be fairly sporadic as there was no training and no standard procedures for managing or documenting projects. The results of the interim project review process were formally recorded only by the researcher. A distinct lack of formal structure required the researcher to acquire relevant knowledge from staff. In eliciting existing knowledge, the researcher relied heavily upon personal networks, particular with ArchSME's quality representative.

In summary, exploitative innovation Hypothesis 2-3 was confirmed, i.e. successful innovation was supported by exploitative capability generated by operational structure capital and relationship capital interaction around a tangible, client-driven business need; whilst unsuccessful innovation was evidenced by disjointed interaction of structure capital and relationship capital at a social level without the benefit of an integrating client-driven business need.

7.4.4   Comment on Hypothesis 2

The outcomes of the testing of the sub-hypotheses are summarised in Table 7.2.

The research results presented a varied picture depending on whether the innovation was explorative or exploitative in nature. For exploitative innovation, Hypothesis 2 was confirmed, i.e. successful exploitative innovation was

Table 7.2 Summary of Hypothesis 2.

Hypothesis	Testing results (confirmed/falsified)	
	Explorative innovation	Exploitative innovation
H 2: capabilities	Falsified	Confirmed
H 2-1: link between HC and RC	Confirmed	Confirmed
H 2-2: link between SC and HC	Falsified	Confirmed
H 2-3: link between RC and SC	Falsified	Confirmed

HC, human capital; SC, structure capital; RC, relationship capital.

characterised by integrated human capital, structure capital and relationship capital (see Sections 7.4.1, 7.4.2 and 7.4.3) around a specific client-driven need; whilst unsuccessful exploitative innovation displaced fragmented human capital, structure capital and relationship capital (see Sections 7.4.1, 7.4.2 and 7.4.3), and did not benefit from a specific client-driven need.

For explorative innovation, Hypothesis 2 appears to be falsified. Successful explorative innovation was characterised by integrated human capital and relationship capital (see Section 7.4.1) around a specific project. The need for integrated structure capital (see Sections 7.4.2 and 7.4.3) was found not to be a prerequisite for successful innovation. This apparent discrepancy that successful innovation can be produced without within strongly coupled formal structure capital is discussed in the meta-hypothesis below.

7.5   Meta-Hypothesis: Knowledge Capital

The meta-hypothesis was set out in Section 3.4.

Meta-hypothesis

A small construction professional practice which generates and integrates relationship capital, structure capital and human capital through exploitative and explorative capabilities will create knowledge capital for successful innovation and sustainable competitive advantage.

7.5.1   Explorative innovation

The knowledge capital for explorative innovation identified in the exploratory phase is the focal or integrating nexus for relationship capital, structure capital and human capital in which innovation takes place (see Section 5.5.4).

In successful explorative innovation, knowledge capital was associated with a combination of 'social' and 'technical' contexts where human capital, structure capital and relationship capital were integrated (see Sections 7.3.1–7.3.3 and 7.4.1–7.4.3), particularly when knowledge capital were applied to operational specific project activity, e.g. new designs (innovation 3) (see Section 5.5.4). The research results indicate that explorative knowledge capital in ArchSME was ultimately through people-to-people dialogue within a social context which brought together relationship capital and human capital. This dialogue was principally supported by social, informal structure capital, for example, face-to-face meetings and telephone conversations targeted at a specific project, through to daily, informal conversations between colleagues and with external supply chain partners, e.g. clients.

In contrast, unsuccessful explorative innovation was seen to be brought about when the knowledge capital was limited to a 'technical' dimension, as it tended to be located at an individual-driven social level and did not lend itself to team-based, socially constructed innovation activity, e.g. new materials (innovation 6) (see Section 5.5.4). The research findings indicate that knowledge capital in unsuccessful explorative innovation was limited to a technical context where human capital and relationship capital were inappropriately integrated.

In summary, for explorative innovation, the meta-hypothesis was confirmed with respect to explorative capability, i.e. successful innovation was characterised by integrated, operational knowledge capital around a project focal; whilst unsuccessful innovation was evidenced by disjointed social knowledge capital around non-specific project context.

7.5.2   Exploitative innovation

The knowledge capital for exploitative innovation identified in the exploratory phase is the same as for explorative innovation, i.e. it is the focal or integrating nexus in which innovation takes place (see Section 5.6.4).

The knowledge capital for successful exploitative innovation was associated with a combination of 'social' and 'technical' contexts where human capital, structure capital and relationship capital were integrated (see Sections 7.3.1–7.3.3 and 7.4.1–7.4.3) at an operational level, e.g. mission statement (innovation 1), IiP (innovation 2) and company restructure (innovation 4) (see Section 5.6.4). For successful exploitative knowledge capital, knowledge workers were connected through social systems (such as being involved in meetings and task forces in the meeting rooms, or in the public house). This enhanced the opportunity for relationship capital and human capital (access information and knowledge among themselves) to interact. The knowledge capital within technical dimension was through electronic documents, handwritten documents, the internet, emails. These technical mechanisms were used to support in human capital and relationship capital interaction. This integrated human capital, structure capital and relationship capital within social and technical contexts converged at a specific project need.

In contrast, in unsuccessful exploitative innovation, knowledge capital targeted at organisational and general client development activity, e.g. seminars (innovation 5) and the Learndirect project (innovation 7) (see Section 5.6.4). These innovations failed as they did not represent tangible, immediate benefits to the firm at a project level. The key findings were confirmed through testing and validating in the action research phase.

The knowledge capital for exploitative innovation within the action research phase was initially stimulated through the 'technical system' through 'encoded' documents and by communication via email, the internet and telephone. This provided the platform to commit the firm's staff to the 'social system' meetings/discussions (see Sections 6.2.2, 6.3.2, 6.4.2 and 6.5.2). This combination of social and technical knowledge capital did not channel into a specific project. The lack of operational project focus was confirmed as key factor for unsuccessful exploitative innovation (see Section 6.6.2).

In summary, for exploitative innovation, the meta-hypothesis was confirmed with respect to exploitative capability, i.e. successful innovation was characterised by integrated, operational knowledge capital around a tangible, client-driven business need; whilst unsuccessful innovation was evidenced by disjointed social knowledge capital around intangible, non-client-driven business need.

7.5.3   Comment on the meta-hypothesis

The meta-hypothesis for exploitative innovation was confirmed, i.e. successful exploitative innovation is generated by exploitative knowledge capital which is a product of appropriately integrated human capital, structure capital and relationship capital with social and technical contexts. In contrast, it was found that successful explorative innovation was not dependent on integrated structure capital; rather, the explorative knowledge capital was principally underpinned by strong relationship capital and human capital interaction around a specific project. This reality is consistent with the central tenet of professional services, namely the co-production of the service between the client and the knowledge worker.

7.6   Summary and Link

This chapter has presented the key findings within the context of the meta-hypothesis and six sub-hypotheses being investigated in the research. The case study results confirmed the prevailing reality that small construction professional practices tend to concentrate their efforts on reactive client-facing, problem-solving innovation (explorative innovation), rather than proactive internal–organisational, general client development innovation (exploitative innovation).

The final chapter summarises this research, draws implications for theory and practice and makes recommendations.

Conclusions

The aim of this chapter is to discuss and summarise the research findings to draw implications for innovation theory and to address the research problem set out in Section 1.2 and research questions articulated in Section 2.7. The structure of this chapter is as follows:

(1) contributions to innovation theory are articulated (Section 8.2);
(2) insights on the research problem based on the results are given (Section 8.3);
(3) the research questions are addressed (Section 8.4);
(4) key limitations and further research are given (Section 8.5);
(5) theoretical and practical implications are noted (Section 8.6); and,
(6) policy implications from this research are provided (Section 8.7).

8.2   Contribution to Innovation Theory

8.2.1   Definition of knowledge-based innovation

The following definition of innovation set out in Section 2.5.5 was found to be useful and valid. Successful knowledge-based innovation is:

> The effective generation and implementation of a new idea which enhances over-all organisational performance, through appropriate exploitative and explorative knowledge capital which develops and integrates relationship capital, structure capital and human capital.

This definition of knowledge-based innovation formed the basis for the knowledge-based innovation concept model.

8.2.2   Knowledge-based innovation concept model

The literature synthesis set out in Chapters 2 and 3 explored the general management and construction specific literature pertaining to innovation in

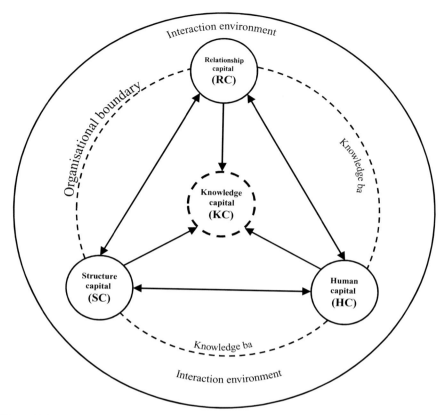

Figure 8.1 Conceptual knowledge-based innovation model for small construction professional practices.

small construction professional practices. The literature review resulted in the knowledge-based innovation concept model set out in Section 3.2 (see Figure 8.1). The literature was found, however, not to extend its consideration to an explicit understanding of how these variables interact with each other (see Section 2.5.5). In developing and testing the conceptual model, this research confirmed the prevailing literature, but in a hitherto inadequately addressed context of small construction professional practices. These variables which make up the model are discussed as follows.

Human capital

The *human capital* (HC) is defined as the capabilities and motivation of individuals within the small construction professional practice, client systems and external supply chain partners to perform productive, professional work in a wide variety of situations (see Section 2.5.4).

The research results confirm the importance of human capital in successful innovation. This is broadly consistent with the prevailing literature which notes that small businesses rely heavily on human capital (e.g. Barber and Manger, 1997). The research findings draw attention to the importance of the company's internal capacity, ability and motivation. This resonates with the literature that stresses that the internal capability to know how to discover, find, filter, gather, store, get access and act on information to optimise performance was particularly important in knowledge-intensive firms (Correia and Sarmento, 2003).

For explorative innovation, the research findings indicate the critical role of staff capacity, ability and motivation. This is consistent with the literature on the role and capabilities of knowledge workers (Quinn *et al.*, 1996). Indeed, it was found that the nature of knowledge-intensive work encouraged staff to be 'self-motivated' in that they are directly responsible for the creation and use of an idea within a project-specific situation. This confirms Maister (1993) who emphasises that professionals are highly self-motivated to perform their own work. This view is extended by Scarborough (1996) and Tampoe (1993) who identify personal growth, operational autonomy and task achievement as key motivators to the knowledge worker.

For exploitative innovation, the research findings indicate the dominant role of senior management, employee participation in decision making and time allocation. First, the role of senior management in exploitative innovation involves the envisioning, creation and application of knowledge. The need for dedicated top management support to motivate senior management sufficiently in driving innovation was emphasised in exploitative innovation. This is in alignment with the literature on small and medium-sized enterprises (SMEs) which notes the significance of the role of the owner–manger in small business (e.g. Carter, 1996; Vyakarnam *et al.*, 1996). Second, the critical role of senior management in providing inspiration for employee participation in decision making was particularly pertinent in exploitative innovation. Without senior management inspiration, employees subsequently become alienated from the innovation implementation process. Finally, the tension between the time and volatility of workload was evident. As a consequence, the need for 'time' resource was addressed in exploitative innovation. This supports Chase (1997) who asserts that lack of time is the one of main barriers to knowledge transfer and innovation.

Structure capital

The *structure capital* (SC) is made up of systems and processes (such as company strategies, machines, tools, work routines and administrative systems) for codifying and storing knowledge from individual, organisation and external supply chain partners (see Section 2.5.4).

The key research findings indicate that principal focus for structure capital in exploitative innovation comprised the administrative systems, the team structure and computer systems; and, in explorative innovation, the team structure and teamwork. The research results reveal that the team structure, teamwork and senior management implementation through the team structure were pertinent in explorative and exploitative innovation. It was found that successful innovation had enduring senior management support from inception through to implementation, and supported by an enabling team structure which stimulated and developed teamwork at an operational level. This is consistent with Starbuck (1992) who notes the importance of social norms of teamwork within knowledge-intensive firms.

The key difference for exploitative innovation (compared to explorative innovation) is the necessity of the formalised systems and documentation systems within the firm. This echoes Blackler (1995) who reports that there is considerable reliance on 'encoded' knowledge by small businesses. The emphasis is on writing and documentation. However, it was found that the outcomes of explorative innovation in terms 'the lesson learned' or 'best practice' did not have sufficient demonstrable benefit or momentum to become embedded in structure capital; rather, the experiential learning stayed with the knowledge worker in a tacit form. This chimes with both the professional service firm literature which stresses that individuals are the principal repositories of firms' competence (Morris and Empson, 1998), and with the small firm literature which emphasises that personal expertise is often not made explicit or codified (Shelton, 2001). The focus on individual, tacit 'repositories' applied within specific projects resonates with the project-based organisation literature which identifies the common dislocation between project-based learning and company-wide learning (e.g. Gann and Salter, 1998). The research findings strongly indicate that in the case of small construction professional practices experiencing rapid growth, the limitation of formalised structures and systems is a restraining force for successful innovation.

In the prevailing literature, 'hard' innovation performance evaluation tools are seen as critical to ensuring improvements in organisation performance (e.g. Ahmed and Zairi, 2000). This research reveals no such clear relationship; rather, innovation is evaluated in a qualitative, ad hoc manner. This arguably is consistent with the co-production nature of professional services but, as has been noted with exploitative innovation, as firms grow in size and complexity, there is an increasing demand for more calibrated, measured approaches to evaluating innovation in order to ensure adequate prioritisation and resource allocation.

Relationship capital

The *relationship capital* (RC) is the network resources of a firm. It results from interactions between individual, organisation and external supplier chain partners, including reputation or image. Relationship capital is the means to leverage human capital (see Section 2.5.4).

The research results confirm that relationship capital provides a critical network of contacts to enable creative action. This is consistent with the literature that relationship capital provides access to knowledge-based resources and is a valuable source of information (e.g. Hendry *et al.*, 1995). Baker (2000), for example, argues that it is not 'what you know', but 'whom you know'. The research findings identify that the key source of relationship capital for explorative innovation was located in internal, client and supplier interactions (see Section 5.5.3); whilst for exploitative innovation was located in business adviser, internal, client and suppliers interactions (see Section 5.6.3). In addition, the research findings reveal 'clients' as being the principal agent in the interaction environment (see Sections 5.5.3 and 5.6.3). The interaction environment is that part of the business environment which firms can interact with, and influence, including 'the task environment' (the environment where this client interaction occurs) and 'the competitive environment' (the environment where other firms which compete with the firm customer and scarce resources) (see Section 2.4).

It was evident that the initial ideas for explorative innovation were to meet specific project needs (client needs); and, the initial ideas for exploitative innovation targeted client-driven business needs. This supports the literature by Schneider and Bowen (1995), who argue that service productivity is, to a significant degree, influenced by the exchange of information and resources between the service provider and client. The importance of client relationships view is emphasised by Tapscott *et al.* (2000, p. 12), who argues that 'the wealth embedded in customer relationships is now more important than the capital contained in land, factories, buildings and even bank accounts.'

The research findings further indicate that supplier interactions are very much meshed with identifying and understanding enabling knowledge, and this process was found to be proactive in nature. This is in alignment with Lee and Yang (2000, p. 787), who argue that the relationship between a corporation and its suppliers is very important and can be regard as an intangible and agile asset of the corporation. Stable and close relationships with suppliers mean that knowledge workers have more access to new, varied knowledge.

Knowledge capital

The *knowledge capital* (*KC*) is the dynamic synthesis of both the 'context' and 'process' of knowledge creation and conversion between individual–organisational–individual knowledge ba spiral, and the 'content' of relationship capital, structure capital and human capital (see Section 2.6). The research results demonstrate knowledge capital to be the focal or integrating social and technical nexus in which innovation takes place.

For explorative and exploitative innovation, knowledge capital in a 'social' context stimulates interaction and collective 'process-orientated' knowledge creation and conversion. It has been widely accepted that organisational knowledge creation is heavily influenced by social processes (Chua, 2002). Nonaka and Takeuchi (1995) argue that the knowledge creation is heavily influenced by social interaction. Communication is the basis constituent in social interaction, according to Luhmann (1990, pp. 86–87): 'without communication there can be no human relations'. A supportive 'social context' within a small construction professional practice can be regarded as a key factor for successful innovation.

The knowledge capital in a 'technical' context supports the search for external knowledge and sharing of 'asset-orientated' knowledge. It takes the form of IT (such as e-mails and the Internet), communication tools (such as

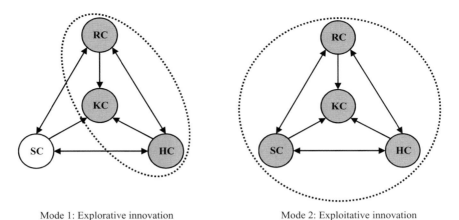

Mode 1: Explorative innovation   Mode 2: Exploitative innovation

Figure 8.2 Types of knowledge-based innovation.

telephones), records (such as handwriting or electronic records) and so on. E-mail, for example, is perceived as being an important IT tool for knowledge-intensive firms (e.g. Robertson *et al.*, 2001).

The research findings note, that through a 'technical' knowledge capital context, knowledge workers easily access explicit knowledge. In contrast, through a 'social' knowledge capital context, knowledge workers share their tacit knowledge, and it is this tacit interaction that is the principal source and driver of successful innovation.

8.2.3   Types of knowledge-based innovation

Two types of knowledge-based innovation identified in Section 5.4 were found to be an appropriate way of categorising the dominant modes of innovation observed in small construction professional practices (see Figure 8.2). They are discussed below:

(1) *Explorative innovation* (*mode 1*) is viewed as innovation which focuses on client facing, project-specific problem solving. Explorative innovation activity heavily relies on the capacity, ability and motivation of staff at an operational level to solve client problems and, in doing so, generates short-term competitive advantage (i.e. project specific). The outcome of this innovation focuses on effective and efficient delivery of services to satisfy current external project needs, but are often not embedded in the organisational structure capital due to management attention and company resources being constantly focused on current or future project-specific considerations (see Section 5.5).

(2) *Exploitative innovation* (*mode 2*) is viewed as an innovation which focuses predominantly on internal organisation and general client development activity which is not project-specific fee earning activity. Exploitative innovation activity heavily relies on the capacity, ability and motivation of senior management at a social level to improve

organisational effectiveness and efficiency to generate sustainable competitive advantage. The distinctive feature of exploitative innovation (compared to explorative innovation) is that new phenomena, systems or structures are securely embedded in the structure capital of the firm (see Section 5.6). This key difference between explorative and exploitative innovation is shown in Figure 8.2.

The concept of explorative and exploitative routines (March, 1991) was introduced in Section 2.5.5. It was noted that explorative routines focused on search, variation, experimentation, flexibility and discovery to create new opportunities and resources to generate sustainable competitive advantage. In contrast, exploitative routines were characterised by refinement and efficiency activities to leverage existing resources to ensure competitive advantage.

The research findings challenge this distinction; indeed, in small construction professional practices, it appears that the focuses of explorative and exploitative routines are reversed. Two modes of knowledge-based innovation have been discerned: explorative innovation and exploitative innovation. Explorative innovation was found to be located in immediate 'new' project domains, and entailed 'search, variation, experimentation, flexibility and discovery' explorative activity to share project-specific problems. In contrast, exploitative innovation concentrated on implementing generic organisational infrastructure (such as ISO 9001 quality management systems) to 'refine' and 'improve the efficiency' of the firm operations to exploit its capability for future activity.

The research findings further provide the characteristic generic and distinctive variables for successful and unsuccessful explorative innovation (see Table 8.1) and for successful and unsuccessful exploitative innovation (see Table 8.2).

Going back to March (1991), the argument put forward is that firms need to have a balance between activities that contribute to exploration of new opportunities, and knowledge and resources and activities that contribute to exploitation of the existing opportunities, knowledge and resources. The balance between exploration and exploitation is key to the understanding of the successful innovating firm. This issue is the focus of the next section.

8.2.4   Definition of a successful knowledge-based innovating firm

The research findings revealed that there was no even balance between, and integration of, explorative and exploitative knowledge capitals. The emphasis was on explorative innovation. Further, the results show that successful explorative innovation appeared to not need integrated structure capital. It was evident, however, that lessons learned from projects were not captured at an exploitative knowledge capital level and fed into current or future projects.

It can be speculated that within small construction professional practices there is too much emphasis on individual learning on the project level (explorative innovation) to be the detriment of the organisational level learning (exploitative innovation). (This deficiency was very much a stimulus for

Table 8.1   Variables for explorative innovation.

Variables	Generic variables	Distinctive variables for successful innovation	Distinctive variables for unsuccessful innovation
Human capital	The capacity, ability and motivation of staff	Social or operational knowledge being applied to meet the project needs	Social knowledge not being applied to meet the project needs
Structure capital	Team structure Teamwork	Team-based ideas Teamwork Senior management involvement through teamwork	Individual-based ideas Individual-based work Senior management not involved in teamwork Limitation of relevant and updated information within the structure
Relationship capital	Operational RC: within internal, client and supplier interactions Social RC: within internal, client and supplier interactions	A combination of operational RC and social RC being applied to meet project needs	Social RC not being applied to meet project needs
Knowledge capital	Social context: company environments (office, meeting room) Technical context: e-mails and the Internet	A combination of social context and technical context	Technical context
Outcome	Effective and efficient delivery of services to satisfy current and/or future project needs	Project performance improvement	Individual performance improvement

the interim project review process innovation described in Chapter 6.) The proposition is shown in Figure 8.3.

At the bottom of the diagram, 'self-contained' projects are shown where often successful explorative innovation has taken place. However, there is no appropriate structure capital in place to encourage the flow of 'project' knowledge capital to 'organisational' knowledge capital at the social level to stimulate exploitative innovation (shown in the top half of the diagram), and vice versa. There is thus no appropriate balance between explorative and exploitative innovation over time.

This lack of integration between explorative and exploitative knowledge capitals, along with the apparent lack of need for integrated structure capital for explorative innovation requires a reconsideration of '*what is successful innovation?*' This emphasis of explorative knowledge capital over exploitative knowledge capital is not sustainable within rapidly growing firms such as ArchSME, as the limitation of structure capital will become increasing

Table 8.2 Variables for exploitative innovation.

Variables	Generic variables	Distinctive variables for successful innovation	Distinctive variables for unsuccessful innovation
Human capital	The capacity, ability and motivation of senior management Employee participation	Top management support Senior management implementation Some employees buy in Training	Top management not supportive Senior management not driving the implementation Lack of time Employees not bought in Inappropriate encouragement Not related to an individual job
Structure capital	The administrative system Team structure Computer systems	Formalised structures and documentation systems Senior management implementation through the team structure	No formalised structures and documentation systems Senior management not driving the implementation through the team structure
Relationship capital	Operational RC: within business adviser, internal, client and supplier interactions Social RC: within internal interactions	A combination of operational RC and social RC being applied to meet project needs	Social RC not being applied to meet project needs
Knowledge capital	Social context: company environments (office and open family culture) Technical context: e-mails and the Internet	A combination of social context and technical context being applied to meet the project needs	A combination of social context and technical context being applied to meet the project needs
Outcome	Organisational effectiveness and efficiency	Organisational performance improvement	Individual performance improvement

evident as a significant restraining force for the effective integration of explorative and exploitative knowledge capitals. (This restraining force has arguably been recognised by ArchSME in its aspiration to become ISO 9001 accredited, and in its decision to adopt the development and use of an interim project review process as the focus of the action research process – see Section 6.2.1.) The ideal balance between explorative and exploitative knowledge capital is shown in Figure 8.4.

However, the lack of balance between explorative and exploitative knowledge capital is not inconsistent with this definition of successful

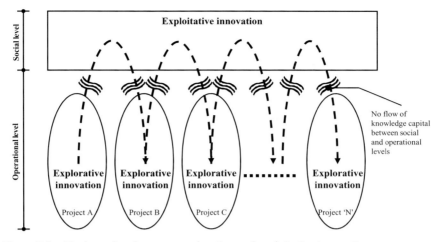

Figure 8.3 The boundary between explorative and exploitative innovation.

knowledge-based innovation proposed in Section 2.5.5. With the benefit of the research findings, it is now evident that this definition is only appropriate for 'an' innovation, but apparently does not adequately address the need for sustainable innovation activity at a firm level, that is, a successful explorative innovation was found to be not creating exploitative knowledge capital to stimulate cumulative learning and innovation across projects over time.

The meta-hypothesis thus ushers in the need to consider not only *'what is successful innovation?'* but *'what is a successful innovating firm?'* The reorientation of the question results in the need to consider the flow of integrated innovation overtime. The following definition of a successful knowledge-based innovating firm is offered to accommodate the time dimension:

> The effective generation and implementation of *a flow of* new ideas which enhance overall organisational performance *over time*, through appropriate exploitative and explorative knowledge capital which develops and integrates relationship capital, structure capital and human capital.

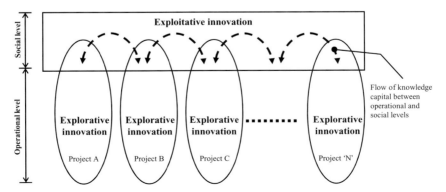

Figure 8.4 Ideal balance between explorative and exploitative knowledge capital.

The time variable brings into focus the development phases of firms as they move from start-up to mature organisations. The focus and process of innovation activity will correspondingly change during the transition. It can be speculated that at the early stages of firms' development, the emphasis is on explorative innovation, but, as firms mature, there is an increasing need to explicitly invest in exploitative innovation. This need was certainly evident in the case study firm.

8.3   Comment on Research Problem

In developing and testing the conceptual knowledge-based innovation model, this research confirms the assertion in Section 1.2 that the prevailing construction guidance for successful innovation is not appropriate for small construction professional practices. Three potential problems of this lack of explicit research into innovation in these types of firms were identified. Each problem is addressed below.

1. Innovation theory tends to be based on manufacturing-based firms; rather than service-based firms in general, and on construction professional practices in particular (see Section 1.2).

The literature review identified that there are significant differences between innovations in manufacturing-based firms and service-based firms (e.g. Miles, 2000). The literature review identified that innovations in the manufacturing sector often emphasise R&D work leading to 'technological' novelties (e.g. Freeman, 1982; Rothwell and Zegfeld, 1982); whilst innovations in the service sector are often based on social networks leading to 'non-technical' innovations (e.g. Kandampully, 2002; Sundbo, 1999).

The research findings confirmed the 'non-technical' emphasis of innovation activity with, for example, effort being allocated to creating a novel mission statement and implementing new IiP management system. Technical innovation was evident in new building designs. This domain of innovation was found to be intrinsically different to manufacturing innovation, however, which creates new products which embody both new component and materials (component innovation) and new linkages between them (architectural innovation) (Henderson and Clark, 1990). In contrast, the technical design innovation was characterised by novel architectural innovation using existing components and materials.

The social characteristics of service innovation compared to manufacturing innovation were also confirmed. Innovation was found to be principally driven by unique co-production of knowledge and innovative solutions between professionals and their clients. This is in contrast to the linear, decoupled nature of manufacturing innovation where 'interaction' is with a homogeneous client 'base'. Further, the literature review identified that innovations in services are often more socially integrated than in manufacturing innovation (Bilderbeek et al., 1994; Sundbo, 1997).

2. Innovation research tends to focus on non-project-based firms in relatively stable supply chains; rather than project-based firms in relatively

unstable supply chains in general, and on construction professional practices in particular (see Section 1.2).

The literature review revealed that there are significant differences between innovations in non-project-based firms and project-based firms (e.g. Gann, 2000; Gann and Salter, 2000). First, the literature review identified that non-project-based firms are better able, through functional hierarchy, to own and maintain innovation compared to project-based firms. Further, the literature review observed that project-based firms are often in loose coupled horizontal transactions between project teams (e.g. Turner and Keegan, 1999). The research findings confirmed that innovation activity, particularly when exploration in nature in the result of co-production with the client. The 'tangible' fruits of innovation activity are 'owned' by the client in the form of an improved building or architectural service. The 'intangible' benefits of innovation do flow to, and accumulate in, individual professionals in the form of tacit knowledge which can be adopted and used in future projects.

Second, the literature review identified that the focus of innovation in non-project-based firms is viewed as improving organisational performance (e.g. Nonaka and Takeuchi, 1995; Young et al., 2001); whilst innovations in project-based firms are often seen as useful, but primarily as costly and dangerous (e.g. Keegan and Turner, 2002). The research findings show that innovations in project-based firms are of benefit for both project and organisational levels (see Sections 5.5.4 and 5.6.4). However, the principal focus was explorative innovation at a project level, as the benefit was seen as immediately and tangibly client focused. This is consistent with the project-based organisation literature which argues that innovation is primarily perused within projects rather than a centralised 'innovation' function (e.g. Becher, 1999; Gann, 1994). In contrast, non-fee earning exploitation innovation was viewed as being of a lower priority, and inherently risks in terms of its opportunity costs of using up finite resource.

3. Innovation research tends to focus on large firms, rather than small firms in general, and on construction professional practices in particular (see Section 1.2).

Four challenges unique to small manufacturing firms were identified (Rothwell and Zegfeld, 1982). They are discussed below.

First, small firms have limited staff capability to undertake R&D compared to large firms. The research findings produced a varied conclusion to this assertion. For explorative innovation, it was found that the firm had sufficient capability to bring about project-based innovation. However, it was evident that the firm lacked capability to undertake non-architectural exploitative innovation; this was apparent in the use of external consultants to develop its quality management systems.

Second, the small firms have scarce time and resources to allocate to external interaction compared to large firms. The research findings did not confirm this argument for explorative innovation as the co-production reality of professional service resulted in continuous interaction with external clients. In contrast, for exploitative innovation, the resource allocation priority to project activity resulted in more limited interaction to absorb external ideas for general organisational development.

Third, small firms are often affected by the excessive influence of senior management. Small firms are often dominated by a single owner or small team who may use inappropriate strategies and skills. The research findings painted a bipolar picture in this regard. At an operational, project level, teams and individual staff were empowered to envision and implement innovation activity with little, if any, intervention from senior management. In contrast, at a social, non-project level, it was found that senior management played a significant gatekeeper role to what innovation activity was prioritised and resourced. This is consistent with the project-based organisation literature which notes that innovation activity is controlled by senior management coalitions (Gann and Salter, 2003). It was evident, however, that the senior management emphasis was on prioritising and resourcing external fee earning project activity.

Finally, small firms can have difficulty in raising finance and maintaining adequate cash flow which can result in limited scope for capital for ongoing innovation in innovation activity compared to large firms. The issue of finance, per se, did not emerge as an enabler or constraint for innovation activity. The co-produced, social nature of project-based innovation made the cost of human capacity the pertinent resource currency. The emphasis on explorative innovation was found to significantly erode the available human resource capacity to progress exploitative innovation.

In summary, the research findings confirmed that the prevailing innovation literature does not adequately capture and explore the unique nuances, characteristics and needs of small construction professional practices.

8.4   Comment on Research Questions

Q1: How do small construction professional practices appropriately develop and manage knowledge interaction activities between individual–organisational–individual (I-O-I) knowledge ba spiral, and how do these arrangements affect innovation performance?

The research findings reveal that successful innovation in small construction professional practices is principally characterised by 'project pull' and 'project push' I-O-I knowledge ba spirals which create dynamic project and/or client-driven knowledge capital. The phenomenon is shown in Figure 8.5.

The left hand side of the figure depicts specific project requirements (either external fee-producing projects or internal client-driven projects) 'pulling', combining and converting, 'organisational knowledge' and 'individual knowledge' to form specific 'project individual knowledge'. Individual project knowledge is integrated and leveraged to create 'project team knowledge' which is appropriately applied to create successful innovation. The feedback I-O-I knowledge ba spiral is complemented (as shown in the right hand side of the figure) by a feedback or 'project push' knowledge ba spiral where new specific 'project team knowledge' feeds back to develop 'project individual knowledge', which, in turn, further enhances 'individual and organisational knowledge'. The tacit, experiential knowledge accumulation and learning is the basis for subsequent cycles of project-based innovation.

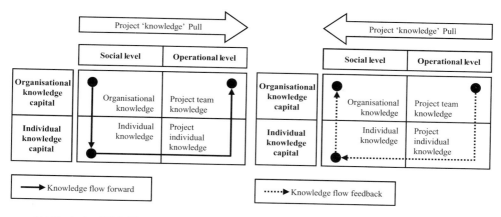

(A) 'Project pull' I-O-I knowledge ba spiral   (B) 'Project pull' I-O-I knowledge ba spiral

Figure 8.5   Successful innovation driven by operational focus.

In contrast, the research findings identify that unsuccessful innovation in small construction professional practices is principally characterised by 'organisation push' of disjointed, unfocused 'social' non-project and/or non-client-driven knowledge capital being 'rejected' by day-to-day project priorities and activities. Without a project focus, innovation fails because the I-O-I knowledge ba spiral does not happen. The phenomenon is shown in Figure 8.6.

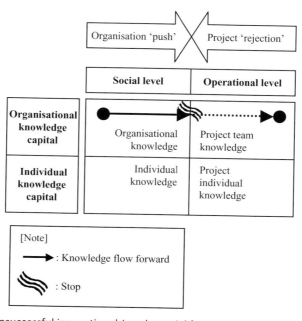

Figure 8.6   Unsuccessful innovation driven by social focus.

Figure 8.6 depicts that there is no specific project needs 'pulling' individual, organisational knowledge together. Rather, generic 'organisational knowledge' is 'pushed' into a project team setting without appropriate filtering and adaptation to meet specific project needs. Further, the 'organisational knowledge' does not benefit from individual knowledge worker championing and tacit understanding. In combination, the 'organisational knowledge' is 'rejected' by the project. As a consequence, the feedback loop through, individual, project and organisational knowledge does not happen.

Q2: How do small construction professional practices appropriately manage and motivate their knowledge workers to create and engage in this development of, and alignment between, the individual–organisational–individual (IOI) knowledge ba spiral?

The research findings identify that successful innovation in small construction professional practices is principally focused on specific project needs and/or client-driven business needs. It was found that the interaction and co-production between the knowledge worker and the client within a 'project setting' is the principal vehicle for managing and motivating knowledge workers. Knowledge workers are intrinsically motivated to undertake interesting knowledge intensive work in their chosen field – in ArchSME's case, to engage with clients to produce high calibre architectural solutions on a project-to-project basis. The research findings indicate that 'senior management commitment' was the key for small construction professional practices to manage and motivate their knowledge workers to create and engage I-O-I knowledge ba spirals (see Figure 8.7).

Senior management commitment to appropriate 'leadership' is necessary to generate an inclusive, galvanising strategic vision which balances and progresses both individual and organisational needs within a project-based setting; and, which empowers knowledge workers to meaningfully 'participate' in the innovation process and to delegate appropriate 'ownership' and

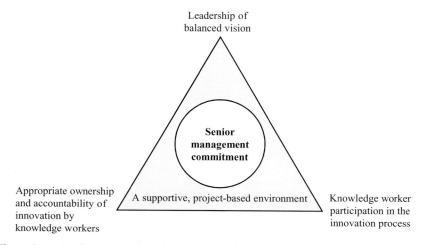

Figure 8.7   An ideal integration of individual and organisational needs.

'accountability' of the innovation to encourage its enduring relevance and success.

For ArchSME, two key practical ways can be identified from the research results to assist in bringing about successful innovation. First, there is a need for senior management to have the capability to manage all aspects of the innovation process. It was evident, for example, in the action research phase, that senior management vision and support was missing at key stages. A contribution to a remedy for this is for senior management to have appropriate education and training in innovation management. Second, effective communication within and between project teams to create and manage innovation activity is essential. It was found that within ArchSME the constant pressure of project delivery hampered this aspect of innovation capability. Senior management should, therefore, establish and adequately resource knowledge sharing meetings which are independent from day-to-day project activity.

8.5   Key Limitations and Future Research

Before discussing the theoretical and practical implications for innovation in small construction professional practices from the research findings, key limitations and associated future research directions are identified. First, a 22-month single case study was used to produce the research findings. The results are thus limited by the degree of representativeness and generalisability of the chosen case study. These limitations have been addressed by a careful sampling strategy to select a representative small construction professional practice based on the size and type of firm. Consequently, we believe that the results can be generalised, with a significant degree of confidence, to the theoretical understanding of innovation within small construction professional practices. Nevertheless, future research would broaden, or appropriately limit, the generalisability of this research by undertaking multiple case studies with a large sample of architectural small construction professional practices and other discipline small construction professional practices (e.g. building surveying and quantity surveying practices). Second, moving on from the limitation above, the focus and process of the innovation process is contingent on its developmental phase: from start-up through to maturity and decline. The case study firm was relatively young. It would therefore be useful if future research would investigate the proposed innovation model and types of innovation by using a longitudinal design over many years in a single case study firm and/or to undertake case study research in a variety of firms at different stages of development. This would provide a richer picture of the innovation process through the developmental life cycle of firms. Finally, there is an emerging literature which is articulating the moderating effects of professional service firm governance structures on organisational priorities and, thus, innovation activity. Empson and Chapman (2006), for example, argue that a shift from an unlimited partnership arrangement to a limited liability partnership results in a corresponding adjustment of strategic priorities from long-term client and professional discipline development to more short-term profit maximisation. The case study firm was a limited liability partnership, which is the dominant governance structure of small construction professional practices in the UK. It would be of interest, however, for future research to apply and evaluate the research findings within a variety of construction professional service firm governance contexts. This strand of enquiry could usefully be extended to consider the impact of new interorganisational governance arrangements on professional service firm innovation activity; for example, partnering and alliance configurations.

8.6   Theoretical and Practical Implications

Despite these limitations, our findings may have several interesting theoretical and practical implications. First, the findings have contributed to our understanding of the distinctive nature of innovation in small construction professional practices by offering an empirically based definition of, and model for, successful innovation. The need for a strategic and systemic investment in, and management of, relationship capital, human capital and structure capital to produce dynamic knowledge capitals which bring about an appropriate balance of explorative and exploitative innovation is emphasised. Second, when compared to innovation in small construction firms (Sexton and Barrett, 2003b), there is a significant difference in emphasis on the:

- pivotal role of the knowledge worker in leading the co-production of innovation with clients, compared to non-knowledge-intensive professional service firms where the owner–managers are the principal innovation champion; and,
- creation of enabling social and structural environments to encourage and capture learning from individual knowledge worker–client interactions so that they can be shared and amplified within an organisational context.

Third, the concept of explorative and exploitative routines has been modified to accommodate the unique characteristics of small construction professional practices. March (1991) notes that explorative routines are focused on search, variation, experimentation, flexibility and discovery to create new opportunities and resources to generate sustainable competitive advantage. In contrast, exploitative routines were characterised by new refinement and efficiency activities to ensure competitive advantage. The first part of this concept resonates with the research findings, with explorative innovation being found to be located in immediate 'new' project domains, and entailed 'search, variation, experimentation, flexibility and discovery' explorative activity to solve project-specific problems. However, the project-specific innovation focus was to generate short-term competitive advantage, that is, a successful project. For sustainable competitive advantage, which departs from March (1991), exploitative innovation was found to be the principal source, with its focus on the design and implementation of generic organisational infrastructure to 'refine' and 'improve the efficiency' of the firm; to 'embed' project-specific knowledge and learning in the structure; and to accumulate, transfer and apply learning within and through projects.

8.7   Policy Implications

The key implications for government and institution policy are twofold. First, as innovation activity is explorative and exploitative in nature, there is a need for policy makers to guide and support the development of new frameworks and tools which emphasise the overtly social nature of innovation in small construction professional practices: it is not a mechanistic, linear process; rather, it is a fluid process where knowledge-based innovation flows from context-specific 'one-off' encounters between knowledge workers and clients at the project level of resolution. The challenge for policy makers, in particular, is to design interventions which encourage these idiosyncratic human-centred flows within the structure of firms and supply chains. Interventions should move away from the current heavy emphasis on codification strategies (such as information technology-based knowledge management systems) which are more attuned to the needs of large construction firms, to a more balanced approach which promotes personalisation strategies to encourage context-specific interaction (such as guidance for the creation and nurturing of communities of practice). Second, within this context, the critical role of senior management in successful innovation within small construction professional practices needs to be better recognised, and this should explicitly shape research priorities and funding. A contribution to a remedy for this, for example, is policy to support the education and training of senior managers in innovation management within a small construction professional practice context.

References

Afuah, A. (1998) *Innovation Management: Strategies, Implementation, and Profit.* Oxford University Press, New York.

Ahmed, R.K. and Zairi, M. (2000) Innovation: A Performance Measurement Perspective. In: J. Tidd (ed.), *From Knowledge Management to Strategic Competence: Measuring Technological, Market and Organisation Innovation.* Imperial College Press, London, pp. 257–294.

Allard-Poesi, F., Drucker-Godard, C. and Ehlinger, S. (2001) Analyzing Representations and Discourse. In: R.A. Thiétart (ed.), *Doing Management Research: A Comprehensive Guide.* Sage, Paris, pp. 351–372.

Alshawi, M. (2007) *Re-thinking IT in Construction and Engineering: Organisational Readiness.* Taylor and Francis, London.

Alvesson, M. (1995) *Management of Knowledge-Intensive Companies.* de Gruyter, Berlin.

Ames, B.C. and Hlavacek, J.D. (1988) *Market Driven Management: Prescription for Survival in a Turbulent World.* Irwin, Homewood, IL.

Amit, R. and Schoemaker, P.J.H. (1993) Strategic Assets and Organizational Rent. *Strategic Management Journal* **14**, 33–46.

Argyris, C. and Schön, D. (1996) *Organization Learning II: Theory, Method and Practice.* Addison-Wesley Publishing, Reading, MA.

Athey, S.E. and Schmutzler, A. (1995) Product and Process Flexibility in an Innovative Environment. *RAND Journal of Economics* **26**(4), 557–574.

Ayas, K. and Zeniuk, N. (2001) Project-Based Learning: Building Communities of Reflective Practitioner. *Management Learning* **32**(1), 61–76.

Baker, W. (2000) *Achieving Success through Social Capital: Tapping the Hidden Resources in Your Personal and Business Networks.* Josey-Bass, San Francisco, CA.

Baleicher, J. (1980) *Contemporary Hermeneutics: Hermeneutics as Method, Philosophy and Critique.* Routledge, London.

Barber, E. and Manger, G. (1997) Improving Management's Valuations of Human Capital in Small Firms. *Journal of Management Development* **16**(7), 457–465.

Barley, S. (1996) Technician in the Workplace: Ethnographic Evidence for Bringing Work into Organization Studies. *Administration Science Quarterly* **41**(1), 146–162.

Barney, J.B. (1991) Firm Resources and Sustained Competitive Advantage. *Journal of Management* **17**(1), 99–120.

Bates, K.B. and Flynn, E.J. (1995) Innovation History and Competitive Advantage: A Resource-Based View Analysis of Manufacturing Technology Innovations. *Academy of Management Journal, Best Paper Proceedings*, pp. 235–239.

Barrett, P. (1993) *Profitable Practice Management: For the Construction Professional.* E & FN Spon Publisher, London.

Barrett, P.S. and Ostergren, K. (1991) The Value of Keypersons in Professional Firms. In: P.S. Barrett and R. Males (eds), *Practice Management: New Perspectives for the Construction Professional*. E & FN Spon Publisher, London, pp. 314–321.

Barrett, P. and Sexton, M. (2006) Innovation in Small, Project-Based Construction Firms. *British Journal of Management* **17**, 331–346.

Barthorpe, S., Chien, H.-J. and Shih, J.K.C. (2003) The Current State of IT or ICT Usage by UK Construction Companies. *International Journal Electronic Business: Special Issue on E-procurement: Myths and Realities* **1**(4), 358–371.

Baumard, P. (2002) Tacit Knowledge in Professional Firms: The Teachings of Firms in Very Puzzling Situations. *Journal of Knowledge Management* **6**(2), 135–151.

Becher, T. (1999) *Professional Practices*. Transaction Publications, London.

Bhatt, G.D. (2002) Management Strategies for Individual Knowledge and Organizational Knowledge. *Journal of Knowledge Management* **6**(1), 31–39.

Bilderbeek, R., Den Hertog, P., Huntink, W., Bouman, M., Kastrinos, N. and Flanagan, K. (1994) *Case Studies in Innovative and Knowledge-Intensive Business Services*. TNO Centre for Technology and Policy Studies, Apeldoorn, The Netherlands.

Blackler, F. (1995) Knowledge, Knowledge Work and Organizations: An Overview and Interpretation. *Organization Studies* **16**(6), 1021–1046.

Blackler, F., Crump, N. and McDonald, S. (1997) Knowledge, Organisation and Competition. In: G. Krogh, J. Roos and D. Kleine (eds), *Knowing in Firms: Understanding, Managing and Measuring Organisational Knowledge*. Sage, London, pp. 253–268.

Bollinger, A.S. and Smith, R.D. (2001) Managing Organizational Knowledge as a Strategic Asset. *Journal Knowledge Management* **5**(1), 8–18.

Bontis, N. (2002) Managing Organizational Knowledge by Diagnosing Intellectual Capital: Framing and Advancing the State of the Field. In: N. Bontis and W.C. Choo (eds), *The Strategic Management of Intellectual Capital and Organizational Knowledge*. Oxford University Press, New York, pp. 621–642.

Boström, E.-O. (1995) Successful Cooperation in Professional Services: What Characteristics Should the Customer Have? *Industrial Marketing Management* **24**, 156–165.

Brandon, P. and Lu, S.L. (eds) (2008) *Clients Driving Innovation*. Wiley-Blackwell Publishing, Oxford.

Breakwell, G.M. (1995) Interviewing. In: G.M. Breakwell, S. Hammond and C. Fife-Shaw (eds), *Research Methods in Psychology*. Sage, London, pp. 230–242.

Brown, J.S. and Duguid, P. (1991) Organizational Learning and Communities of Practice: Toward a Unifying View of Working, Learning and Innovation. In: M.D. Cohen and L.S. Spruoll (eds), *Organizational Learning*. Sage, London, pp. 59–82.

Burgelman, R., Maidique, M. and Wheelwright, S. (1996) *Strategic Management of Technology and Innovation*, 2nd edn. Irwin, Homewood.

Carrillo, P. (2004) Managing Knowledge: Lessons from the Oil and Gas Sector. *Construction Management and Economics* **22**(6), 631–642.

Carter, S. (1996) Small Business Marketing. In: M. Warner (ed.), *International Encyclopaedia of Business and Management*, Vol. 5. Routledge, London, pp. 4502–4509.

Chaharbaghi, K. and Lynch, R. (1999) Sustainable Competitive Advantage: Towards a Dynamic Resource-Based Strategy. *Management Decision* **37**(1), 45–50.

Chase, R.L. (1997) The Knowledge-Based Organisation: An International Survey. *Journal of Knowledge Management* **1**(1), 38–49.

Chaston, I., Badger, B., Mangles, T. and Sadler-Smith, E. (2002) Knowledge-Based Services and the Internet: An Investigation of Small UK Accountancy Practices. *Journal of Small Business and Enterprise Development* **9**(1), 49–60.

Chaston, I., Badger, B. and Sadler-Smith, E. (1999) Organizational Learning: Research Issues and Application in SME Sector Firms. *International Journal Entrepreneurial Behaviour & Research* **5**(4), 191–203.

Chua, A. (2002) The Influence of Social Interaction on Knowledge Creation. *Journal of Intellectual Capital* **3**(4), 375–392.

Churchill, N.C. and Lewis, V.L. (1983) The Five Stages of Small Business Growth. *Harvard Business Review* **61**(3), 30–50.

Civil Engineering Research Foundation (CERF) (1997) Commercializing Infrastructure Technologies: A Handbook for Innovators. CERF Report# 97-5028, Washington, DC.

Civil Engineering Research Foundation (CERF) (2000) Guidelines for Moving Innovations into Practice. *Working Draft Guidelines for the CERF International Symposium and Innovative Technology Tradeshow*, 14–17 August, CERF, Washington, DC.

Clarke, R. (1993) *Industrial Economics*. Blackwell Publishing, Oxford.

Clippinger, J.H. (1995) Visualization of Knowledge: Building and Using Intangible Assets Digitally. *Planning Review* **23**(6), 28–31.

Cohen, M.D. (1991) Individual Learning and Organisational Routine: Emerging Connections. *Organization Science* **2**(1), 135–139.

Cohen, D. (1998) Toward a Knowledge Context: Report on the First Annual UC Berkeley Forum on Knowledge and the Firm. *California Management Review: Special Issue on Knowledge and the Firm* **40**(3), 22–39.

Cohen, D. and Prusak, L. (2001) *In Good Company: How Social Capital Makes Organizations Work*. Harvard Business School Press, Boston, MA.

Coleman, J.S. (1988) Social Capital in the Creation of Human Capital. *American Journal Sociology* **94**, S95–S120.

Collins, H.M. (1993) The Structure of Knowledge. *Social Research* **60**(1), 95–116.

Collis, D.J. (1994) Research Note: How Valuable Are Organisational Capabilities. *Strategic Management Journal* **15**, 143–152.

Commission of The European Communities (2007) Towards a European Strategy in Support of Innovation in Services: Challenges and Key Issues for Future Actions, Brussels, 27.07.2007, SEC(2007) 1059. Available at http://ec.europa.eu/enterprise/innovation/doc/com_2007_1059_en.pdf (accessed 18 July 2008).

Construction Industry Council (CIC) (2003) Survey of UK Construction Professional Services 2001/2002, January, CIC, London. Available at http://www.cic.org.uk/activities/FinalReport03.pdf (accessed 7 May 2008).

Construction Industry Council (CIC) (2008) *Survey of UK Construction Professional Services 2005/2006*. CIC, London.

Correia, A.M.R. and Sarmento, A. (2003) Knowledge Management: Key Competences and Skills for Innovation and Competitiveness. *Proceeding of the Technology & Human Resource Management International Conference*, 18–23 May, Ceram Sophia Antipolis, France.

Cunningham, I. (1988) Interactive Holistic Research: Researching Self-Managed Learning. In: P. Reason (ed.), *Human Inquiry in Action: Developments in New Paradigm Research*. Sage, London, pp. 163–181.

Das, T. (1983) Qualitative Research in Organisational Behaviour. *Journal of Management Studies* **20**(3), 301–314.

Davenport, T.H., De Long, D.W. and Beers, M.C. (1998) Successful Knowledge Management Projects. *Sloan Management Review* **39**(2), 43–57.

Davenport, T.H. and Prusak, L. (1998) *Working Knowledge: How Organizations Manage What They Know*. Harvard University Press, Boston, MA.

Day, E. and Barksdale, H.C. (1992) How Firms Select Professional Services. *Industrial Marketing Management* **21**, 85–91.

De Long, D.W. and Fahey, L. (2000) Diagnosing Cultural Barriers to Knowledge Management. *Academy of Management Executive* **14**(4), 113–127.

Denzin, N.K. (1978) *The Research Act: A Theoretical Introduction to Sociological Methods*, 2nd edn. McGraw-Hill, London.

Department for Business, Enterprise and Regulatory Reform (BERR) (2006) SME STATISTICS 2006 – Table 3 UK Industry Summary and Table 5 UK Divisions, BERR: Enterprise Directorate Analytical Unit. Available at http://stats.berr.gov.uk/ed/sme/ (accessed 7 February 2008).

Department for Business, Enterprise and Regulatory Reform (BERR) and HM Treasury (HMT) (2007) Productivity in the UK 7: Securing Long-Term Prosperity, BERR and HMT, November, London. Available at http://www.berr.gov.uk/files/file42710.pdf (accessed 29 July 2008).

Department of Trade and Industry (DTI) (1998) *Our Competitive Future: Building the Knowledge Driven Economy* HMSO, London.

Department of Trade and Industry (DTI) (2002) *UK Competitiveness Indicators*, 2nd edn. HMSO, London. Available at: http://217.154.27.195/competitiveness/index.htm (accessed 29 July 2008).

Department of Trade and Industry (DTI) (2003) *Innovation Report – Competing in the Global Economy: The Innovation Challenge.* HMSO, London. Available at http://www.berr.gov.uk/files/file12093.pdf (accessed 7 July 2008).

Despres, C. and Hiltrop, J. (1995) Human Resource Management in the Knowledge Age: Current Practice and Perspectives on the Future. *Journal of Employee Relations* **17**(1), 9–23.

Dougherty, V. (1999) Knowledge Is about People, Not Databases. *Industrial and Commercial Training* **51**(7), 262–266.

Drucker, P.F. (1993) *Post-Capitalist Society*. Butterworth-Heinemann, London.

Drucker, P. (1997) The Future That Has Already Happened. *Harvard Business Review* **75**(5), 20–24.

Eden, C. (1992) On the Nature of Cognitive Maps. *Journal of Management Studies* **29**, 261–265.

Egan, J. (1998) *Rethinking Construction: Report of the Construction Task Force to the Deputy Prime Minister, John Prescott, on the Scope For Improving the Quality and Efficiency of UK Construction, DETR.* HMSO, London.

Egbu, C.O. (1999) The Role of Knowledge Management and Innovation in Improving Construction Competitiveness. *Building Technology and Management Journal* **25**, 1–10.

Egbu, C.O., Henry, J., Kaye, G.R., Quintas, P., Schumacher, T.R. and Young, B.A. (1998) Managing Organisational Innovations in Construction. *Proceedings of the Association of Researchers in Construction Management Fourteenth Annual Conference*, 9–11 September, University of Reading, Reading.

Egbu, C.O., Sturges, J. and Gorse, C. (2000) Communication of Knowledge for Innovation within Projects and across Organisational Boundaries. *Congress 2000, 15 World Congress on Project Management*, 22–25 May, Royal Lancaster Hotel, London, UK.

Empson, L. and Chapman, C. (2006) Partnership versus corporation: Implications of Alternative Forms of Governance in Professional Service Firms. *Research in the Sociology of Organizations* **24**, 145–76.

European Commission (EC) (1995) Green Paper on Innovation, December, EC, DG XIII.

European Commission (EC) (2000) European Competitive Report 2000, EC, Belgium.

European Commission (EC) (2003) Commission Recommendation of 6 May 2003 Concerning the Definition of Micro, Small and Medium-Sized Enterprises, Official Journal the European Union, OJL 124 of 20.5.2003, pp. 36–41.

European Commission (EC) (2007) Green Paper – The European Research Area: New Perspectives, Brussels. Available at http://www.grad.ac.uk/downloads/documents/Reports/European%20reports/EC%20green%20paper%20on%20ERA%20Apr%2007.doc (accessed 7 July 2008).

Fairclough, J. (2002) *Rethinking Construction Industry Innovation and Research: A Review of Government R & D Policies and Practices.* Department of Trade and Industry, HMSO, London. Available at: http://217.154.27.195/competitiveness5old/Pch1/index.htm (accessed 7 July 2008).

Fiol, M. and Lyles, M. (1985) Organisational Learning. *Academy of Management Review* 10(4), 803–813.

Fontana, A. and Frey, J.H. (2000) Interviewing: The Art of Science. In: N.K. Denzin and Y.S. Lincoln (eds), *Handbook of Qualitative Research.* Sage, Thousand Oaks, London, pp. 361–376.

Forss, K., Cracknell, B. and Samset, K. (1994) Can Evaluation Help an Organisation to Learn? *Evaluation Review* 18(5), 574–591.

Freeman, C. (1982) *The Economics of Industrial Innovation.* Penguin, Harmondsworth, London.

Galliers, R.D. and Newell, S. (2000) Back to the Future: From Knowledge Management to Data Management. Working Paper No. 92, Department of Information Systems, London School of Economics and Political Science, London, UK.

Gann, D. (1994) Innovation in the Construction Sector. In: M. Dodgson and R. Rothwell (eds), *The Handbook of Industrial Innovation.* Edward Elgar, Aldershot, pp. 202–212.

Gann, D. (2000) *Building Innovation: Complex Constructs in a Changing World.* Thomas Telford, London.

Gann, D.M. and Salter, A.J. (1998) Learning and Innovation Management in Project-Based, Service-Enhanced Firms. *International Journal of Innovation Management* 2(4), 431–454.

Gann, D.M. and Salter, A.L. (2000) Innovation in Project-Based, Service-Enhanced Firms: The Construction of Complex Products and Systems. *Research Policy* 29, 955–972.

Gann, D. and Salter, A. (2003) Project Baronies: Growth and Governance in the Project-Based Firm. *Proceedings of the DRUID Summer Conference: Creating, Sharing and Transferring Knowledge: The Role of Geography,* 12–14 June, Institutions and Organizations, Copenhagen, Denmark.

Ghemawat, P. and Costa, J. (1993) The Organizational Tension between Static and Dynamic Efficiency. *Strategic Management Journal* 14, 59–73.

Girmscheid, G. and Hartmann, A. (2002) Innovation in Construction: The View of Client. *Proceedings of the 10th International Symposium: Construction Innovation and Global Competitiveness: The Organization and Management of Construction,* CIB W55–W65, 9–13 September, Cincinnati, Ohio, USA, Vol. 1, pp. 29–43.

Glückler, J. and Armbrüster, T. (2003) Bridging Uncertainty in Management Consulting: The Mechanisms of Trust and Networked Reputation. *Organization Studies* 7(4), 269–297.

Grant, R.M. (1995) *Contemporary Strategy Analysis: Concepts, Techniques, Applications,* 2nd edn. Blackwell Publishing, Oxford.

Grant, R.M. (1996a) Prospering in Dynamically Competitive Environments: Organizational Capability as Knowledge Integration. *Organization Science* 7(4), 375–387.

Grant, R.M. (1996b) Toward a Knowledge-Based Theory of the Firm. *Strategic Management Journal* **17**, 109–122.

Green, S.D., Larsen, G.D. and Kao, C. (2008) Competitive Strategy Revisited: Contested Concepts and Dynamic Capabilities. *Construction Management and Economics* **26**(1), 63–78.

Greenwood, R. and Suddaby, R. (eds) (2006) *Professional Service Firms, Research in the Sociology of Organizations*, Vol. 24. JAI Press, London.

Grönroos, C. (2000) *Service Management and Marketing: A Customer Relationship Management Approach*, 2nd edn. Wiley, Chichester.

Gupta, A.K., Smith, K.G. and Shalley, C.E. (2006) The Interplay between Exploration and Exploitation. *The Academy of Management Journal* **49**(4), 693–708.

Gurran, J. and Blackburn, R.A. (2001) *Researching the Small Enterprise*, 1st edn. Sage, London.

Hadjimanolis, A. (2000) A Resource-Based View of Innovativeness in Small Firms. *Technology Analysis and Strategic Management* **12**(2), 263–281.

Håkanson, H. (1989) *Corporate Technological Behaviour: Corporation and Networks*. Printer, London.

Hansson, J. (2002) Management of Knowledge Transfer in Knowledge Service Firms. *Proceedings of the EURAM 2002: Innovative Research in Management*, 9–11 May, Stockholm, Sweden.

Harty, C., Goodier, C.I., Soetanto, R., Austin, S., Dainty, A.R.J. and Price, A.D.F. (2007) The Futures of Construction: A Critical Review of Construction Future Studies. *Construction Management and Economics* **25**(5), 477–493.

Henderson, R. and Clark, K.B. (1990) Architectural Innovation: The Manufacturing of Existing Product Technologies and the Failure of Established Firms. *Administrative Science Quarterly* **35**, 9–30.

Hendry, C., Arthur, M. and Jones, A. (1995) *Strategy through People: Adaptation and Learning in the Small-Medium Enterprise*. Routledge, London.

Hildreth, P. and Kimble, C. (2004) *Knowledge Networks: Innovation through Communities of Practice*. Idea Group Publishing, Hershey, PA.

Hill, C.J. and Neely, S.E. (1988) Differences in the Consumer Decision Process for Professional vs. Generic Services. *Journal of Services Marketing* **2**(1), 17–23.

Holmqvist, M. (2003) A Dynamic Model of Intra -and Interorganizational Learning. *Organization Studies* **24**, 93–121.

Hussey, J. and Hussey, R. (1997) *Business Research: A Practical Guide for Undergraduate and Postgraduate Students*. Macmillan, London.

Ibarra, H. (1993) Network Centrality, Power, and Innovation Involvement: Determinants of Technical and Administrative Roles. *Academy of Management Journal* **36**(3), 471–501.

Imai, K. (1992) The Japanese Pattern of Innovation and Its Evaluation. In: N. Rosenberg, R. Handau and D. Mowery. (eds), *Technology and The Wealth of Nations*. Stanford Press, Stanford.

Itami, H. (1987) *Mobilizing Invisible Assets*. Harvard University Press, Cambridge, MA.

Jick, T.D. (1979) Mixing Qualitative and Quantitative Methods: Triangulation in Action. *Administrative Science Quarterly* **24**, 602–611.

Kandampully, J. (2002) Innovation as the Core Competency of Service Organisation: The Role of Technology, Knowledge and Networks. *European Journal of Innovation Management* **5**(1), 18–26.

Keegan, A. and Turner, J.K. (2002) The Management of Innovation in Project-Based Firms. *Long Range Planning* **35**, 367–388.

Knapp, E.M. (1998) Knowledge Management. *Business and Economic Review* **44**(4), 3–6.

Knock, N. and McQueen, R. (1998) Knowledge and Information Communication within Organization: An Analysis of Core, Support and Improvement Process. *Knowledge & Process Management* 5(1), 29–40.

Knott, A.M. (2002) Exploration and Exploitation as Complements. In: C.W. Choo and N. Bontis. (eds), *The Strategic Management of Intellectual Capital and Organizational Knowledge: A Collection of Readings.* Oxford University Press, New York, pp. 299–358.

Kotler, P. (1980a) *Principles of Marketing.* Prentice-Hall International, Englewood Cliffs, NJ.

Kotler, P. (1980b) *Marketing Management: Analysis Planning and Control.* Prentice-Hall, Englewood Cliffs, NJ.

Kotler, P. and Bloom, P.N. (1984) *Marketing Professional Services.* Prentice-Hall, Englewood Cliffs, NJ.

Kululanga, G.K. and McCaffer, R. (2001) Measuring Knowledge Management for Construction Organizations. *Engineering, Construction and Architectural Management* 8(5-6), 346–354.

Lamb, C.E. (2003) An Assessment of the Impact of Investors in People on Architectural Practice. Unpublished dissertation, April, Master of Business Administration, Manchester Metropolitan University, Manchester, UK.

Landry, R., Amara, N. and Lamari, M. (2002) Does Social Capital Determine Innovation? To What Extent? *Technological Forecasting and Social Change* 69(7), 681–701.

Latham, M. (1994) *Constructing the Team.* HMSO, London.

Lee, C. and Yang, J. (2000) Knowledge Value Chain. *Journal of Management Development* 19(9), 783–793.

Leedy, P.D. (1988) *Practical Research: Planning and Design.* Macmillan, New York.

Leonard-Barton, D. (1990) A Dual Methodology For Case Studies: Synergistic Use of a Longitudinal Single Site with Replicated Multiple Sites. *Organization Science* 1(3), 248–266.

Leonard-Barton, D. (1992) Core Capability and Core Rigidities: A Paradox in Managing New Product Development. *Strategic Management Journal* 13, 111–125.

Loosemore, M., Dainty, A. and Lingard, H. (2003) *Human Resource Management in Construction Projects: Strategic and Operational Approaches.* Spon Press, London.

Løwendahl, B.R. (2000) *Strategic Management of Professional Service Firms*, 2nd edn. Handeshøjskolens Forlag, Denmark.

Lu, S. and Sexton, M.G. (2004) Appropriate Research Design for Investigating Innovation in Small Knowledge-Intensive Professional Service Firms. *Proceedings of the ARCOM 20th Annual Conference and Annual General Meeting*, Heriot Watt University, Edinburgh, UK, 1–3 September, pp. 733–739.

Lu, S. and Sexton, M. (2006) Innovation in Small Construction Knowledge-Intensive Professional Service Firms: A Case Study of an Architectural Practice. *Construction Management and Economics* 24(12), 1269–1282.

Luhmann, N. (1990) *Essays on Self-Reference.* Columbia University Press, New York.

MacDuffie, J.P. (1995) Human Resource Bundles and Manufacturing Performance: Organizational Logic and Flexible Production Systems in the World Auto Industry. *Industrial and Labor Relations Review* 48(2), 197–221.

Maister, D.H. (1993) *Management the Professional Service Firm.* Simon and Schuster, New York.

Mansfield, E. (1991) *Microeconomics.* Norton, New York.

March, J.G. (1991) Exploration and Exploitation in Organizational Learning. *Organization Science* 2(1), 71–87.

March, J.G. and Levinthal, D.A. (1999) The Myopia of Learning. In: J.G. March. (ed.), *The Pursuit of Organisational Intelligence*. Blackwell, Oxford, pp. 191–222.

Miles, I. (2000) *Services Innovation: Coming of Age in the Knowledge-Based Economy*. International Journal of Innovation Management 4(4), 371–389.

Miles, M.B. and Huberman, A.M. (1994) *Qualitative Data Analysis: A Sourcebook*. Sage, Thousand Oaks, CA.

Miozzo, M. and Ivory, C. (1998) *Innovation in Construction: A Case Study of Small and Medium-Sized Construction Firms in the North West of England*. Manchester School of Management, UMIST, Manchester.

Morris, T. and Empson, L. (1998) Organisation and Expertise: An Exploration of Knowledge Bases and the Management of Accounting and Consulting Firms. *Accounting, Organizations and Society* 23(5–6), 609–624.

Mukherjee, A., Mitchell, W. and Talbot, F.B. (2000) The Impact of New Manufacturing Technologies and Strategically Flexible Production. *Journal of Operations Management* 18, 139–169.

Muller, E. (2001) *Innovation Interaction between Knowledge-Intensive Business Services and Small and Medium-Size Enterprises: An Analysis in Terms of Evolution*. Knowledge and Territories, Physica-Verlag, Heidelberg, Germany.

Nanda, A. (1996) *Resources, Capabilities and Competences*. Sage, London.

Neilson, R.E. (1997) *Collaborative Technologies and Organizational Learning*. Idea Group Publishing, London.

Nelson, P.R.. (ed.) (1993) *National Innovation Systems: A Comparative Analysis*. Oxford University Press, Oxford.

Nonaka, I. and Konno, N. (1998) The Concept of 'Ba': Building a Foundation for Knowledge Creation. *California Management Review* 40(3), 40–54.

Nonaka, I. and Takeuchi, H. (1995) *The Knowledge-Creating Company: How Japanese Companies Create the Dynamics of Innovation*. Oxford University Press, New York.

Nonaka, I., Toyama, R. and Konno, N. (2001) SECI, Ba and Leadership: A Unified Model of Dynamic Knowledge Creation. In: K. Nonaka and D.J. Teece. (eds), *Managing Industrial Knowledge: Creation, Transfer and Utilization*. Sage, London, pp. 13–43.

Nooteboom, B. (1994) Innovation and Diffusion in Small Firms: Theory and Evidence. *Small Business Economics* 6, 327–347.

Nordhaug, B. (1993) *Human Capital in Organization: Competence, Training and Learning*. Scandinavian University Press, Oslo.

Ojasalo, J. (1999) Quality Dynamics in Professional Services. PhD Thesis, Swedish School of Economics and Business Administration, Ekonomi och samhälle No. 76, Helsinki, Finland.

Oppenheim, A.N. (1992) *Questionnaire Design, Interviewing and Attitude Measurement*. Printer, London.

Organisation for Economic Co-operation and Development (OECD) (2006) *Factbook 2006: Economic, Environmental and Social Statistics*. OECD, Paris.

Orr, J.E. (1990) Sharing Knowledge Celebrating Identify: Community Memory in a Service Culture. In: D. Middleton and D. Edwards. (eds), *Collective Remembering*. Sage, Newburg Park, pp. 169–189.

Page, M., Limeneh, M., Pearson, S. and Pryke, S. (1999) Understanding Innovation in Construction Professional Service Firms: A Study of Quantity Surveying Firms. *Proceedings of the RICS Construction and Building Research Conference (COBRA): The Challenge of Change: Construction and Building for the New Millennium*, University of Salford, 1–2 September, Vol. 1, pp. 122–130.

Penrose, E.T. (1959) *The Theory of the Growth of the Firm*. Wiley, New York.

Peters, T. (1994) *The Tom Peters Seminar: Crazy Times Call for Crazy Organizations*. Macmillan, London.

Polanyi, M. (1962) *Personal Knowledge: Towards a Post Critical Philosophy*. Routledge and Kegan Paul, London.

Polanyi, M. (1967) *The Tacit Dimension*. Routledge and Kegan Paul, London.

Porter, M.E. (1980) *Competitive Strategy: Techniques for Analyzing Industries and Competitors*. The Free Press, New York.

Porter, M.E. (1985) *Competitive Advantage: Creating and Sustaining Superior Performance*. The Free Press, New York.

Porter, M.E. (1990) *The Competitive Advantage of Nations*. The Free Press, New York.

Prenciple, A. and Tell, F. (2001) Inter-Project Learning: Processes and Outcomes of Knowledge Codification in Project-Based Firms. Research Policy **30**(9), 1373–1394.

Quinn, J.B. (1992) *Intelligent Enterprise: A Knowledge and Service Based Paradigm for Industry*. Free Press, New York.

Quinn, J., Anderson, P. and Finkelstein, S. (1996) Managing Professional Intellect: Making the Most of the Best. *Harvard Business Review* **74**(2), 71–80.

Rabey, G. (2000) Whither HR? Don't People Matter Anymore. *Industrial and Commercial Training* **32**(1), 19–23.

Raich, M. (2002) HRM in the Knowledge-Based Economy: Is There an Afterlife? *Journal of European Industrial Training* **26**(6), 269–273.

Richards, L. (1999) *Using NVivo in Qualitative Research*. Sage, London.

Robert Huggins Associates (2006) Trends and Drivers of Change in the European Knowledge-Intensive Business Services Sector: Mapping Report. European Foundation for the Improvement of Living and Working Conditions. Available at http://www.eurofound.europa.eu/pubdocs/2006/40/en/1/ef0640en.pdf (accessed 18 July 2008).

Robertson, M., Sørensen, C. and Swan, J. (2001) Survival of the Leanest: Intensive Knowledge Work and Groupware Adaptation. *Information Technology & People* **14**(4), 334–352.

Rogers, E.M. (1983) *Diffusion of Innovations*, 3rd edn. The Free Press, New York.

Rothwell, R. (1989) Small Firms, Innovation and Industrial Change. *Small Business Economic* **1**, 51–64.

Rothwell, R. and Dodgson, M. (1994) Innovation and Firm Size. In: M. Dodgson and R. Rothwell. (eds), *The Handbook of Industrial Innovation*. Edward Elgar, Aldershot, Hants, pp. 310–324.

Rothwell, R. and Zegfeld, W. (1982) *Innovation and the Small and Medium Sized Firm*. Printer, London.

Royer, I. and Zarlowski, P. (2001) Sampling. In: R.A. Thiétart. (ed.), *Doing Management Research: A Comprehensive Guide*. Sage, Paris, pp. 147–171.

Scarborough, H.. (ed.) (1996) *The Management of Expertise*. Macmillan, London.

Schneider, B. and Bowen, D. (1995) *Winning the Service Game*. Harvard Business School Press, Boston, MA.

Scott, M.C. (1998) *The Intellect Industry: Profiting and Learning from Professional Service Firms*. John Wiley, Chichester.

Seaden, G., Gouolla, M., Doutriaux, J. and Nash, J. (2001) Analysis of the Survey on Innovation, Advanced Technologies and Practices in the Construction and Related Industry 1999. The Institute for Research in Construction of the National Research Council of Canada and by the Science, Innovation and Electronic Information Division of Statistics Canada, 88F0017MIE, No. 10, January, Canada.

Sexton, M.G. and Barrett, P.S. (2003a) A Literature Synthesis of Innovation in Small Construction Firms: Insights, Ambiguities and Questions. *Construction Management and Economics: Special Issue on Innovation in Construction* **21**, 613–622.

Sexton, M.G. and Barrett, P.S. (2003b) Appropriate Innovation in Small Construction Firms. *Construction Management and Economics: Special Issue on Innovation in Construction* **21**, 623–633.

Sexton, M.G. and Barrett, P.S. (2004) The Role of Technology Transfer in Innovation within Small Construction Firms. *Engineering, Construction & Architecture Management* **11**(5), 342–348.

Sexton, M.G. and Barrett, P.S. (2005) Performance-Based Building and Innovation: Balancing Client and Industry Needs. *Building Research and Information* **33**(2), 142–148.

Simon, H. (1957) *Administrative Behaviour*, 2nd edn. Macmillan, New York.

Simon, J.L. and Burstein, P. (1985) *Basic Research Methods in Social Science*, 3rd edn. Random House, London.

Shapero, A. (1985) *Managing Professional People: Understanding Creative Performance*. Free Press, New York.

Shelton, R. (2001) Helping a Small Business Owner to Share Knowledge. *Human Resource Development International* **4**(4), 429–450.

Slaughter, S.E. (1998) Models of Construction Innovation. *Journal of Construction Engineering and Management, ASCE.* **124**(3), 226–231.

Stake, R. (1994) Case Studies. In: N.K. Denzin and Y.S. Lincoln. (eds), *Handbook of Qualitative Research*. Sage, London, pp. 236–247.

Stalk, G., Evans, P. and Shulman, L.E. (1992) Competing on Capabilities, The New Rules of Corporate Strategy. *Harvard Business Review* **70**, 57–69.

Starbuck, W.H. (1992) Learning by Knowledge-Intensive Firms. *Journal of Management Studies* **29**(6), 713–740.

Stewart, T.A. (1997) *Intellectual Capital: The New Wealth of Organisations*. Doubleday/Currency, New York.

Storey, D.J. (1994) *Understanding the Small Business Sector*. Routledge, London.

Sundbo, J. (1997) Management of Innovation in Services. *The Service Industries Journal* **17**(3), 432–455.

Sundbo, J. (1999) Balancing Empowerment. *Technovation* **16**(8), 397–409.

Susman, G.I. (1983) Action Research: A Sociotechnical Systems Perspective. In: G. Morgan. (ed.), Beyond Method: Strategies for Social Research. Sage, London, pp. 95–113.

Sveiby, K.E. (1997) *The New Organizational Wealth: Managing and Measuring Knowledge-Based Assets*. Berrett-Koehler, San Francisco, CA.

Sverlinger, P.M. (2000) Managing Knowledge in Professional Service Organisation: Technical Consultants Serving the Construction Industry. PhD Thesis, Department of Service Management, Chalmers University of Technology, Göteborg, Sweden.

Synder, H. and Pierce, J. (2002) Intellectual Capital. In: B. Cronin. (ed.), *Annual Review of Information Science and Technology*, Vol. 36. Information Today, Medford, NJ, pp. 467–500.

Takeuchi, H. (2001) Towards a Universal Management of the Concept of Knowledge. In: K. Nonaka and D.J. Teece. (eds), *Managing Industrial Knowledge: Creation, Transfer and Utilization*. Sage, London, pp. 315–329.

Tampoe, M. (1993) Motivating Knowledge Workers: The Challenge for the 1990s. *Long Range Planning* **26**(3), 49–55.

Tapscott, D., Ticoll, D. and Lowy, A. (2000) *Digital Capital: Harnessing the Power of Business Webs*. Harvard Business School Press, Boston, MA.

Teece, D.J., Pisano, G. and Shuen, A. (1997) Dynamic Capabilities and Strategic Management. In: N.J. Foss. (ed.), *Resources, Firms and Strategies*. Oxford University Press, New York, pp. 268–287.

The Society of British Aerospace Companies (SBAC) (2002) *High Performance Work Organization in UK Aerospace: The SBAC Human Capital Audit 2002*. SBAC, London.

Thomas, D.R.E. (1975) Strategy Is Different in Service Business. *Harvard Business Review* **56**, 158–165.

Thompson, J.D. (1967) *Organizations in Action*. McGraw-Hill, New York.

Turner, J.R. and Keegan, A.E. (1999) The Versatile Project-Based Organisation: Governance and Operational Control. *The European Management Journal* **17**(3), 296–309.

van de Ven, A.H., Polley, D., Garud, R. and Venkataraman, S. (1999) *The Innovation Journey*. Oxford University Press, New York.

Vogt, W.P. (1993) *Dictionary of Statistics and Methodology*. Sage, Newbury Park.

von Hippel, E. (1988) *The Sources of Innovation*. Oxford University Press, New York.

Vyakarnam S., Jacobs, R. and Handelberg, J. (1996) Building and Managing Relationships: The Core Competence of Rapid Growth Business. *Proceedings of the 19th ISBA National Small Firms Policy and Research Conference-Enterprising Futures*, UCE Business School, Birmingham, Vol. **1**, pp. 661–683.

Walsh, J.P. and Ungson, G.R. (1991) Organizational Memory. *Academy of Management Review* **16**, 57–91.

Warglien, M. (2000) *The Evolution of Competences in a Population of Projects: A Case Study*. University of Venice Publication, Venice, Italy.

Wasko, M. and Faraj, S. (2000) It Is What One Does: Why People Participate and Help Others in Electronic Communities of Practice. *Journal of Strategic Information Systems* **9**(2–3), 155–173.

Weber, R.P. (1985) *Basic Content Analysis*. Sage, London.

Wharton, A. (2004) Constrinnonet Final Report: Innovative Issues, Successful Practice & Improvements. European Commission, Brussels.

Wheatley, E.W. (1983). *Marketing Professional Services*. Prentice-Hall, Englewood Cliffs, NJ.

Wilson, A. (1972) *The Marketing of Professional Services*. McGraw-Hill Book Company, London.

Wilson, T.L. (1997) Segment Profitability of the US Business Services Sector: Some Reflections on Theory and Practice. *International Journal of Service Industry Management* **8**(5), 398–413.

Winch, G. and Schneider, E. (1993) Managing the Knowledge-Based Organisation: The Case of Architectural Practice. *Journal of Management Studies* **30**(6), 923–937.

Woiceshyn, J. and Falkenberg, L. (2008) Value Creation in Knowledge in Knowledge-Based Firms: Aligning Problems and Resources. *Academy of Management Perspectives* **22**(2), 85–99.

Wood, P. (2001) Regional Innovation and Business Service. *Scott Policy Seminar*, May, NIERC, Belfast.

Wood, S. and de Menezes, L. (1998) High Commitment Management in the UK: Evidence from the Workplace Industrial Relations Survey and Employers' Manpower and Skills Practices Survey. *Human Relations* **51**(4), 485–515.

Yamazaki, Y. and Ueda, Y. (2003) Technology and Knowledge Fusions toward Construction Innovation. *Proceedings of the Joint International Symposium of CIB Working Commissions, Department of Building, National University of Singapore*, Singapore, 22–24 October, Vol. **1**, pp. 40–53.

Yin, R.K. (1994) *Case Study Research: Design and Methods, Applied Social Research Methods Series*, 2nd edn. Sage, Newbury Park, CA.

Yin, R.K. (2003) *Case Study Research: Design and Methods, Applied Social Research Methods Series*, 3rd edn. Sage, London.

Yli-Renko, H., Autio, E. and Sapienza, H.J. (2001) Social Capital, Knowledge Acquisition, and Knowledge Exploitation in Young Technology-Based Firms. *Strategic Management Journal* 22, 587–613.

Young, G.L., Charns, M.P. and Shortell, S.M. (2001) Top Manager and Network Effects on the Adoption of Innovative Management Practices: A Study of TQM in a Public Hospital System. *Strategic Management Journal* 22, 935–951.

Appendices

Appendix A: List of Company Documentation

Number	Description
1	Company handbook
2	ArchSME quality manual
3	CAD handbook
4	Examples of job forms
5	Examples of drawing issue sheets
6	Examples of site record sheets
7	Examples of snagging sheets
8	Examples of nonconformity reports
9	Examples of nonconformity spreadsheets
10	Examples of audit schedules
11	Examples of audit check lists
12	Examples of audit reports
13	Examples of telephone conversations records
14	Examples of induction records
15	Examples of employee continuous professional development records
16	Examples of client correspondence
17	Examples of ArchSME correspondence

Appendix B: Cooperation Proposal

	Phase 1: Analysis	Phase 2: Evaluation workshop	Phase 3: Innovation

Objective	• **General fact finding**
	  – Understand general information about the firm and its employees, its key area of business, market(s) which operating
	• **Successful / unsuccessful innovation within the firm**
	  – Understand drivers, enabler and barriers for the firm to successful / unsuccessful innovation
		• Feedback findings to the firm about its innovation performance and potential areas for improvement
		• Identify innovation activity which company would benefit from University of Salford input
			• Work in collaboration to bring about a successful innovation activity

Information gathering approach	• Interviews • Company documentation	• Workshop	• Interviews (involving in appropriate meeting, etc.) • Observation • Company documentation • Questionnaire survey
People	• 1–2 senior management • 1–2 project architects/managers • 1–2 architect assistants/senior technicians/technicians	• 5–6 key people	• All relevant employees within the firm
Resource implications	• Up to 90 minutes for each interview • Assess to company documents	• Conference room (can be at University of Salford if required) • Up to 3 hours	• Assess to company documents • Involvement in appropriate company activity • Up to 6 months
Time	01/11/2003–30/11/2003	01/12/2003–15/12/2003	01/01/2004–30/06/2004

⇨ **Deliverable: General Finding Report**
• Identification the firm's innovation performance and potential areas for improvement

⇨ **Deliverable: Workshop Report**
• Understanding the firm's current innovation performance
• Identification potential areas for future innovation performance improvement

⇨ **Deliverable: Final Report**
• Feedback on strength and weakness of innovation activity, and identification of potential areas for improvement

Appendix C: Interview Cooperation Proposal

Innovation Survey 2003 in ArchSME

- **Aim of this survey** is to help your company to innovate successfully and profitably.
- **Purpose of this survey** is to gather the information and experiences of your company in successful/unsuccessful innovation activities.
- **Interview and workshop plan**

		Phase 1: Interviews	**Phase 2: Workshop**
Objective		• Understand general information about the firm and you • Understand drivers, enabler and barriers for the firm to successful / unsuccessful innovation from a wide range of business operations	• Evaluation interview findings to the firm about its innovation performance and potential areas for improvement
Information gathering approach		• Face-to-face interview • An audio recorder will be used through your interview	• Group discussion • An audio recorder will be used through the workshop
Resource implications	**Time**	• Up to 90 minutes	• Up to 3 hours
	Documentation	• Assess to company documentation where appropriate related to innovation activities	• Assess to company documentation where appropriate related to innovation activities
Attendees		• You	• You and other interviewees
Venue		• ArchSME	• Conference room in the ArchSME
Duration		17/11/2003–19/12/2003	01/01/2004–16/01/2004

Deliverable: the word-processed document

- The content of your interview will be transcribed verbatim into the word-processed document.
- The word-processed document will be send to you in order to check accuracy.

Deliverable: Workshop Report

Confidentiality

Although this survey requests your name and other specific information, this is only for our purposes and will not be passed on to third parties or attributed directly in any public way.

Appendix D: Interview Protocol

Ref:	
Date:	

Salford University - SCPM

Innovation in SMEs in the Construction Industry

THE 2003 INTERVIEW PROTOCOL

INTRODUCTION

The interview is aimed for better understanding successful innovations in small knowledge-intensive professional service firms (SKIPSFs).

By knowledge-based innovation we mean "*the effective generation and implementation of a new idea, through appropriate development of, and conversion between, relationship capital, structural capital and human capital, to create knowledge capital, which enhances overall organisational performance.*"

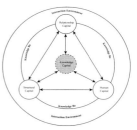

SECTION 1 is designed to collect background information about you, the company and its clients.

SECTION 2 aims to understand in the way of your firm appropriate innovation and identify valuable resources and competencies in innovation activities.

You can go outside the boundaries of these questions to illustrate significant points you feel are important.

Thank you for your time and support. Transcript will be sent to you for you to confirm that I have understood what you have said correctly.

| Confidential | **Ref:** | |
| | **Date:** | |

<table>
<tr><td colspan="2" align="center">SECTION 1: BACKGROUND</td></tr>
</table>

We are aware that much of the information we are asking you to give about your company may be commercially, or in other ways, highly sensitive. However, we assure you that all responses will be treated as STRICTLY CONFIDENTIAL and will be used for research purposes only. You or your company will not be identified or named in any publication arising from the research without your permission. Only aggregated data will be used.

A. About you

Name: _____ Age: _____ No. of years with this company: _____

Tel. No.: _____ Fax. No.: _____ E-mail: _____

Your role/activity: _____

Q1: Please tick "one box" to describe your position within your company:

☐ Top level management

☐ Middle level management

☐ First level supervisor

☐ Professional employee without supervisory role

☐ Other (Please specify) _____

Q2: Please tick "the relevant boxes" to describe your formal qualification:

☐ Graduate (☐ Cognate / ☐ Non cognate)

☐ Trainee members of professional institutions (Please specify) _____

☐ Fully qualified members of professional institutions (Please specify) _____

☐ Trained to Higher National Certificate or Higher National Diploma

☐ Other (Please specify) _____

Q3: Please tick "the relevant boxes" to describe your pervious company's status:

No	No. of years with it	Main products/services	Type	Size
Company 1			☐ Public ☐ Private	☐ Micro organisation (1–10 employees) ☐ Small organisation (11–49 employees) ☐ Medium organisation (50–250 employees) ☐ Large organisation (more than 251 employees)
Company 2			☐ Public ☐ Private	☐ Micro organisation (1–10 employees) ☐ Small organisation (11–49 employees) ☐ Medium organisation (50–250 employees) ☐ Large organisation (more than 251 employees)
Company 3			☐ Public ☐ Private	☐ Micro organisation (1–10 employees) ☐ Small organisation (11–49 employees) ☐ Medium organisation (50–250 employees) ☐ Large organisation (more than 251 employees)

Confidential

Ref:	
Date:	

B. About your company (only one interviewee answer)

1. General corporate information

Company name and address: _____

_____

Tel. No.: _____ Fax. No.: _____ Website address: _____

The business first established in_____(year)

Company history: _____

2. Company profile

Q1: Please tick "the relevant boxes" to describe your company's main activities and state approximately the percentage of its workload:

☐ Multi disciplinary (_____ %)   ☐ Civil and Structural Engineering (_____ %)
☐ Architectural (_____ %)   ☐ Planning (_____ %)
☐ Surveying (_____ %)   ☐ Project management (_____ %)
☐ Quantity surveying (_____ %)   ☐ Management consultancy (not project related) (___ %)
☐ Building services engineering (_____ %) ☐ Other (Please specify) _____ (_____ %)

Q2: Please tick "one box" to describe your company's status:

☐ Public (public limited company with public investors)   ☐ Proprietary (with owner managers)
☐ Subsidiary (controlled by a parent company)   ☐ Private (owners separate from management)
☐ Joint Venture   ☐ Other (Please specify) _____

Q3: How many people does your firm currently employ?

No. of employees   ☐☐☐

No. of full-time employees   ☐☐☐   No. of part-time employees   ☐☐☐

No. of fixed term or contact employees   ☐☐☐   No. of temporary employees   ☐☐☐

Q4: How many employees do you have 12 months ago?

☐☐☐

Q5: How many employees do you anticipate having in 12 months time?

☐☐☐

Ref:	
Date:	

Q6: What was your company's approximate turnover (£) for the last five financial years?

2003
£ [ ] ,000   2002
£ [ ] ,000   2001
£ [ ] ,000

2000
£ [ ] ,000   1999
£ [ ] ,000

Q7: What was your company's approximate pre-tax profit (£) for the last five financial years?

2003
£ [ ] ,000   2002
£ [ ] ,000   2001
£ [ ] ,000

2000
£ [ ] ,000   1999
£ [ ] ,000

Q8: Does your company currently hold any patents/intellectual property rights? If yes, what are they?

3. Profile of clients

Q1: Please identify your principal clients:

No.	Client name and percentage of its workload	Type	Size	Why do you feel this client comes to you
Client 1		☐ Public ☐ Private	☐ Micro organisation (1–10 employees) ☐ Small organisation (11–49 employees) ☐ Medium organisation (50–250 employees) ☐ Large organisation (more than 251 employees)	
Client 2		☐ Public ☐ Private	☐ Micro organisation (1–10 employees) ☐ Small organisation (11–49 employees) ☐ Medium organisation (50–250 employees) ☐ Large organisation (more than 251 employees)	
Client 3		☐ Public ☐ Private	☐ Micro organisation (1–10 employees) ☐ Small organisation (11–49 employees) ☐ Medium organisation (50–250 employees) ☐ Large organisation (more than 251 employees)	

Confidential	**Ref:**	
	Date:	

SECTION 2: KNOWLEDGE-BASED INNOVATION DETAILS

(Note: Knowledge-based innovation is defined as *"the effective generation and implementation of a new idea, through appropriate development of, and conversion between, relationship capital, structural capital and human capital, to create knowledge capital, which enhances overall organisational performance."*)

A. Definition

Q1: How would you describe the term of 'knowledge' in the individual, organisational and supply chain level?

Description:	Discussion:
Tacit ☐☐☐☐☐ Explicit	

Q2: How would you describe the term of 'innovation' in the individual, organisational and supply chain level?

Description:	Discussion:
Tacit ☐☐☐☐☐ Explicit	

Confidential	Ref:	
	Date:	

B. Company environment

Q1: Does your firm have any formal, written business strategy? If yes, how does it operate?

Notes:	Description	Discussion
• What is your business strategy? • How is the business strategy developed into the firm? • How is the business strategy implemented into the firm? • What are the role of directors, staff, clients and communication?	Informal □□□□□ Formal	

Q2: Does your firm have any innovation strategy? If yes, how does it operate?

Notes:	Description	Discussion
• What is your innovation strategy (e.g. IT, technology, communication, rewards/employee incentive etc.)? • How is the innovation strategy developed into the firm? • How is the innovation strategy implemented into the firm? • What are the role of directors, staff, clients and communication?	Informal □□□□□ Formal	

Q3: How does your company foster relationships (including those with its workforce, suppliers and clients) to encourage innovation activities?

Notes:	Description	Discussion	Implied enablers /obstacles
• What activities were carried out to foster relationships? • How were these activities carried out? • Who carried out these activities?	Informal □□□□□ Formal		

Confidential

	Ref:	
	Date:	

Q4: How does your firm develop the ability and motivation of its staff to bring about innovation?

Notes:	Description	Discussion	Implied enablers /obstacles
• What activities were carried out to develop the ability and motivation of staff? • How were these activities carried out? • Who carried out these activities?	 Informal ☐☐☐☐☐ Formal		

Q5: What structures and processes within your firm encourage/discourage innovation activities?

Notes:	Description	Discussion	Implied enablers /obstacles
• What are the firm's structures and processes? • What activities were carried out to encourage/discourage innovation activities? • How were these activities carried out? • Who carried out these activities?	 Informal ☐☐☐☐☐ Formal		

Q6: What knowledge management activity is in place to encourage knowledge sharing around innovations to take place?

Notes:	Description	Discussion	Implied enablers /obstacles
• What is the firm's knowledge management activity? • What activities were carried out to encourage knowledge sharing? • How were these activities carried out? • Who carried out these activities?	 Informal ☐☐☐☐☐ Formal		

Confidential	Ref:	
	Date:	

C. Successful Innovation

C1. Identify successful innovations

Q1: Please identify 'ONE' significant firm-generated successful innovation over the last two years. (Only senior manager answer)

No.	Chosen innovation	Rate level of influence	Strength	Why considered significant
Innovation 1		Senior management driven	+ ☐☐☐☐☐ -	
		Construction client driven	+ ☐☐☐☐☐ -	
		Legislation driven	+ ☐☐☐☐☐ -	
		Competitor driven	+ ☐☐☐☐☐ -	
		Your customer(s) driven	+ ☐☐☐☐☐ -	
		Your supplier(s) driven	+ ☐☐☐☐☐ -	

Q2: Please identify 'ONE' significant firm-generated successful innovation over the last two years. (All interviewees answer except senior manager)

No.	Chosen innovation	Rate level of influence	Strength	Why considered significant
Innovation 2		Senior management driven	+ ☐☐☐☐☐ -	
		Construction client driven	+ ☐☐☐☐☐ -	
		Legislation driven	+ ☐☐☐☐☐ -	
		Competitor driven	+ ☐☐☐☐☐ -	
		Your customer(s) driven	+ ☐☐☐☐☐ -	
		Your supplier(s) driven	+ ☐☐☐☐☐ -	

Confidential

Ref:	
Date:	

C2. Successful Innovation 1 _____ (Senior manager identified)

1. Generate a new idea

Q1: Where did the initial idea(s) come from?

Notes:	Description	Discussion	Implied enablers / obstacles
• What information sources were used (e.g. clients, suppliers, colleagues, reports etc.)? • What activities were carried out to scan for/collect information? • How were these activities carried out? • Who carried out these activities?	Tacit ☐☐☐☐☐ Explicit		

How would rate the following characteristics of the information sources:	Strength	Details
Accessibility	+ ☐☐☐☐☐ -	
Cost	+ ☐☐☐☐☐ -	
Fit	+ ☐☐☐☐☐ -	

Q2: How was the idea adopted?

Notes:	Description	Discussion	Implied enablers /obstacles
• What activities were carried out to innovate from this information (e.g. evaluation etc.)? • How were these activities carried out? • Who carried out these activities?	Informal ☐☐☐☐☐ Formal		

How would rate the following characteristics of your company in the idea adopting phase:	Strength	Details
Strategic focus	+ ☐☐☐☐☐ -	
Communication: Internal	+ ☐☐☐☐☐ -	
Communication: External	+ ☐☐☐☐☐ -	
Resourcing	+ ☐☐☐☐☐ -	
Authority	+ ☐☐☐☐☐ -	
Knowledge management	+ ☐☐☐☐☐ -	

Confidential		Ref:	
		Date:	

2. Implement a new idea

Q1: How was the idea exploited?

Notes:	Description	Discussion	Implied enablers / obstacles
• How was the innovation commercialise/utilised? • What activities were carried out to commercialise/utilise this innovation? • How were these activities carried out? • Who carried out these activities?	 Active ☐☐☐☐☐ Passive		

How would rate the following characteristics of your company in the exploitation phase:	Strength	Details
Supply chain focus / commitment	+ ☐☐☐☐☐ -	
Marketing	+ ☐☐☐☐☐ -	
Resourcing	+ ☐☐☐☐☐ -	
Authority	+ ☐☐☐☐☐ -	

3. Company support

Q1: How was this innovation supported by your firm's relationships?

Notes:	Description	Discussion	Enablers / obstacles
• What relationships were used to support this innovation? • What activities were carried out to support this innovation? • How were these activities carried out? • Who carried out these activities?	 Informal ☐☐☐☐☐ Formal		

Confidential	Ref:	
	Date:	

Q2: How did your firm develop the ability and motivation of its staff to support this innovation?

Notes:	Description	Discussion	Enablers / obstacles
• What activities were used to develop the ability and motivation of staff in order to support this innovation? • What activities were carried out to support this innovation? • How were these activities carried out? • Who carried out these activities?	Informal ☐☐☐☐☐ Formal		

Q3: How did structures and processes within your firm support this innovation?

Notes:	Description	Discussion	Enablers / obstacles
• What structures and processes were used to support this innovation? • What activities were carried out to support this innovation? • How were these activities carried out? • Who carried out these activities?	Informal ☐☐☐☐☐ Formal		

Q4: How was this innovation supported by knowledge management activity?

Notes:	Description	Discussion	Enablers / obstacles
• What knowledge management activity was used to support this innovation? • What activities were carried out to support this innovation? • How were these activities carried out? • Who carried out these activities?	Informal ☐☐☐☐☐ Formal		

Confidential		Ref:	
		Date:	

4. Innovation performance measurement/indicators

Q1: What were the impacts from this innovation?

Notes:	Description	Discussion	Implied enablers / obstacles
• What were the expected/unexpected positive impacts from this innovation? • What were the expected/unexpected negative impacts from this innovation?			
	Tacit ☐☐☐☐☐ Explicit		

Q2: How did you measure this innovation performance?

Notes:	Description	Discussion	Implied enablers / obstacles
• What measurement/indicators (e.g. stakeholder attitudes, business results, etc.) were used to measure this innovation performance? • What activities were carried out to measure this innovation performance? • How were these activities carried out? • Who carried out these activities?			
	Informal ☐☐☐☐☐ Formal		

Q3: How do you intend to further develop/exploit the benefits from this innovation?

Notes:	Description	Discussion	Implied enablers / obstacles
• What activities were carried out to develop/exploit the benefits from this innovation? • How were these activities carried out? • Who carried out these activities?			
	Informal ☐☐☐☐☐ Formal		

Confidential

Ref:	
Date:	

C3. Successful Innovation 2 _____ (You identified)

1. Generate a new idea

Q1: Where did the initial idea(s) come from?

Notes:	Description	Discussion	Implied enablers / obstacles
• What information sources were used (e.g. clients, suppliers, colleagues, reports etc.)? • What activities were carried out to scan for/collect information? • How were these activities carried out? • Who carried out these activities?	 Tacit ☐☐☐☐☐ Explicit		

How would rate the following characteristics of the information sources:	Strength	Details
Accessibility	+ ☐☐☐☐☐ -	
Cost	+ ☐☐☐☐☐ -	
Fit	+ ☐☐☐☐☐ -	

Q2: How was the idea adopted?

Notes:	Description	Discussion	Implied enablers / obstacles
• What activities were carried out to innovate from this information (e.g. evaluation etc.)? • How were these activities carried out? • Who carried out these activities?	 Informal ☐☐☐☐☐ Formal		

How would rate the following characteristics of your company in the idea adopting phase:	Strength	Details
Strategic focus	+ ☐☐☐☐☐ -	
Communication: Internal	+ ☐☐☐☐☐ -	
Communication: External	+ ☐☐☐☐☐ -	
Resourcing	+ ☐☐☐☐☐ -	
Authority	+ ☐☐☐☐☐ -	
Knowledge management	+ ☐☐☐☐☐ -	

Confidential				**Ref:**	
				Date:	

2. Implement a new idea

Q1: How was the idea exploited?

Notes:	Description	Discussion	Implied enablers / obstacles
• How was the innovation commercialise/utilised? • What activities were carried out to commercialise/utilise this innovation? • How were these activities carried out? • Who carried out these activities?	 Active ◻◻◻◻◻ Passive		

How would rate the following characteristics of your company in the exploitation phase:	Strength	Details
Supply chain focus / commitment	+ ◻◻◻◻◻ -	
Marketing	+ ◻◻◻◻◻ -	
Resourcing	+ ◻◻◻◻◻ -	
Authority	+ ◻◻◻◻◻ -	

3. Company support

Q1: How was this innovation supported by your firm's relationships?

Notes:	Description	Discussion	Enablers / obstacles
• What relationships were used to support this innovation? • What activities were carried out to support this innovation? • How were these activities carried out? • Who carried out these activities?	 Informal ◻◻◻◻◻ Formal		

Confidential

Ref:	
Date:	

Q2: How did your firm develop the ability and motivation of its staff to support this innovation?

Notes:	Description	Discussion	Enablers / obstacles
• What activities were used to develop the ability and motivation of staff in order to support this innovation? • What activities were carried out to support this innovation? • How were these activities carried out? • Who carried out these activities?	Informal ☐☐☐☐☐ Formal		

Q3: How did structures and processes within your firm support this innovation?

Notes:	Description	Discussion	Enablers / obstacles
• What structures and processes were used to support this innovation? • What activities were carried out to support this innovation? • How were these activities carried out? • Who carried out these activities?	Informal ☐☐☐☐☐ Formal		

Q4: How was this innovation supported by knowledge management activity?

Notes:	Description	Discussion	Enablers / obstacles
• What knowledge management activity was used to support this innovation? • What activities were carried out to support this innovation? • How were these activities carried out? • Who carried out these activities?	Informal ☐☐☐☐☐ Formal		

Confidential				Ref:	
				Date:	

4. Innovation performance measurement/indicators

Q1: What were the impacts from this innovation?

Notes:	Description	Discussion	Implied enablers / obstacles
• What were the expected/unexpected positive impacts from this innovation? • What were the expected/unexpected negative impacts from this innovation?	Tacit ☐☐☐☐☐ Explicit		

Q2: How did you measure this innovation performance?

Notes:	Description	Discussion	Implied enablers / obstacles
• What measurement/indicators (e.g. stakeholder attitudes, business results, etc.) were used to measure this innovation performance? • What activities were carried out to measure this innovation performance? • How were these activities carried out? • Who carried out these activities?	Informal ☐☐☐☐☐ Formal		

Q3: How do you intend to further develop/exploit the benefits from this innovation?

Notes:	Description	Discussion	Implied enablers / obstacles
• What activities were carried out to develop/exploit the benefits from this innovation? • How were these activities carried out? • Who carried out these activities?	Informal ☐☐☐☐☐ Formal		

Confidential

Ref:	
Date:	

D. Unsuccessful Innovation

D1. Identify unsuccessful innovations

Q1: Please identify 'ONE' potentially significant innovations over the last two years which failed. (Only senior manager answer)

No.	Chosen innovation	Rate level of influence	Strength	Why considered significant
Innovation 1		Senior management driven	+ ☐☐☐☐☐ -	
		Construction client driven	+ ☐☐☐☐☐ -	
		Legislation driven	+ ☐☐☐☐☐ -	
		Competitor driven	+ ☐☐☐☐☐ -	
		Your customer(s) driven	+ ☐☐☐☐☐ -	
		Your supplier(s) driven	+ ☐☐☐☐☐ -	

Q2: Please identify 'ONE' potentially significant innovations over the last two years which failed. (All interviewees answer except senior manager)

No.	Chosen innovation	Rate level of influence	Strength	Why considered significant
Innovation 2		Senior management driven	+ ☐☐☐☐☐ -	
		Construction client driven	+ ☐☐☐☐☐ -	
		Legislation driven	+ ☐☐☐☐☐ -	
		Competitor driven	+ ☐☐☐☐☐ -	
		Your customer(s) driven	+ ☐☐☐☐☐ -	
		Your supplier(s) driven	+ ☐☐☐☐☐ -	

	Ref:	
Confidential	Date:	

D2. Unsuccessful Innovation 1 _____ (Senior manager identified)

1. Generate a new idea

Q1: Where did the initial idea(s) come from?

Notes:	Description	Discussion	Implied enablers / obstacles
• What information sources were used (e.g. clients, suppliers, colleagues, reports etc.)? • What activities were carried out to scan for/collect information? • How were these activities carried out? • Who carried out these activities?	Tacit ☐☐☐☐☐ Explicit		

How would rate the following characteristics of the information sources:	Strength	Details
Accessibility	+ ☐☐☐☐☐ -	
Cost	+ ☐☐☐☐☐ -	
Fit	+ ☐☐☐☐☐ -	

Q2: How was the idea adopted?

Notes:	Description	Discussion	Implied enablers / obstacles
• What activities were carried out to innovate from this information (e.g. evaluation etc.)? • How were these activities carried out? • Who carried out these activities?	Informal ☐☐☐☐☐ Formal		

How would rate the following characteristics of your company in the idea adopting phase:	Strength	Details
Strategic focus	+ ☐☐☐☐☐ -	
Communication: Internal	+ ☐☐☐☐☐ -	
Communication: External	+ ☐☐☐☐☐ -	
Resourcing	+ ☐☐☐☐☐ -	
Authority	+ ☐☐☐☐☐ -	
Knowledge management	+ ☐☐☐☐☐ -	

Ref:	
Date:	

2. Implement a new idea

Q1: How was the idea exploited?

Notes:	Description	Discussion	Implied enablers / obstacles
• How was the innovation commercialise/utilised? • What activities were carried out to commercialise/utilise this innovation? • How were these activities carried out? • Who carried out these activities?	 Active ☐☐☐☐☐ Passive		

How would rate the following characteristics of your company in the exploitation phase:	Strength	Details
Supply chain focus / commitment	+ ☐☐☐☐☐ -	
Marketing	+ ☐☐☐☐☐ -	
Resourcing	+ ☐☐☐☐☐ -	
Authority	+ ☐☐☐☐☐ -	

3. Company support

Q1: How was this innovation supported by your firm's relationships?

Notes:	Description	Discussion	Enablers / obstacles
• What relationships were used to support this innovation? • What activities were carried out to support this innovation? • How were these activities carried out? • Who carried out these activities?	 Informal ☐☐☐☐☐ Formal		

Confidential		Ref:	
		Date:	

Q2: How did your firm develop the ability and motivation of its staff to support this innovation?

Notes:	Description	Discussion	Enablers / obstacles
• What activities were used to develop the ability and motivation of staff in order to support this innovation? • What activities were carried out to support this innovation? • How were these activities carried out? • Who carried out these activities?	Informal ☐☐☐☐☐ Formal		

Q3: How did structures and processes within your firm support this innovation?

Notes:	Description	Discussion	Enablers / obstacles
• What structures and processes were used to support this innovation? • What activities were carried out to support this innovation? • How were these activities carried out? • Who carried out these activities?	Informal ☐☐☐☐☐ Formal		

Q4: How was this innovation supported by knowledge management activity?

Notes:	Description	Discussion	Enablers / obstacles
• What knowledge management activity was used to support this innovation? • What activities were carried out to support this innovation? • How were these activities carried out? • Who carried out these activities?	Informal ☐☐☐☐☐ Formal		

Confidential	Ref:	
	Date:	

4. Innovation performance measurement/indicators

Q1: What were the impacts from this innovation?

Notes:	Description	Discussion	Implied enablers / obstacles
• What were the expected/unexpected positive impacts from this innovation? • What were the expected/unexpected negative impacts from this innovation?	Tacit ☐☐☐☐☐ Explicit		

Q2: How did you measure this innovation performance?

Notes:	Description	Discussion	Implied enablers / obstacles
• What measurement/indicators (e.g. stakeholder attitudes, business results, etc.) were used to measure this innovation performance? • What activities were carried out to measure this innovation performance? • How were these activities carried out? • Who carried out these activities?	Informal ☐☐☐☐☐ Formal		

Q3: How do you intend to further develop/exploit the benefits from this innovation?

Notes:	Description	Discussion	Implied enablers / obstacles
• What activities were carried out to develop/exploit the benefits from this innovation? • How were these activities carried out? • Who carried out these activities?	Informal ☐☐☐☐☐ Formal		

Confidential	Ref:	
	Date:	

D3. Unsuccessful Innovation 2 _____(You identified)

1. Generate a new idea

Q1: Where did the initial idea(s) come from?

Notes:	Description	Discussion	Implied enablers / obstacles
• What information sources were used (e.g. clients, suppliers, colleagues, reports etc.)? • What activities were carried out to scan for/collect information? • How were these activities carried out? • Who carried out these activities? Tacit ☐☐☐☐☐ Explicit			

How would rate the following of the information sources:	Strength	Details
Accessibility	+ ☐☐☐☐☐ -	
Cost	+ ☐☐☐☐☐ -	
Fit	+ ☐☐☐☐☐ -	

Q2: How was the idea adopted?

Notes:	Description	Discussion	Implied enablers / obstacles
• What activities were carried out to innovate from this information (e.g. evaluation etc.)? • How were these activities carried out? • Who carried out these activities? Informal ☐☐☐☐☐ Formal			

How would rate the following characteristics of your company in the idea adopting phase:	Strength	Details
Strategic focus	+ ☐☐☐☐☐ -	
Communication: Internal	+ ☐☐☐☐☐ -	
Communication: External	+ ☐☐☐☐☐ -	
Resourcing	+ ☐☐☐☐☐ -	
Authority	+ ☐☐☐☐☐ -	
Knowledge management	+ ☐☐☐☐☐ -	

Confidential

Ref:	
Date:	

2. Implement a new idea

Q1: How was the idea exploited?

Notes:	Description	Discussion	Implied enablers / obstacles
• How was the innovation commercialise/utilised? • What activities were carried out to commercialise/utilise this innovation? • How were these activities carried out? • Who carried out these activities? Active ☐☐☐☐☐ Passive			

How would rate the following characteristics of your company in the exploitation phase:	Strength	Details
Supply chain focus / commitment	+ ☐☐☐☐☐ -	
Marketing	+ ☐☐☐☐☐ -	
Resourcing	+ ☐☐☐☐☐ -	
Authority	+ ☐☐☐☐☐ -	

3. Company support

Q1: How was this innovation supported by your firm's relationships?

Notes:	Description	Discussion	Enablers / obstacles
• What relationships were used to support this innovation? • What activities were carried out to support this innovation? • How were these activities carried out? • Who carried out these activities? Informal ☐☐☐☐☐ Formal			

Confidential		Ref:	
		Date:	

Q2: How did your firm develop the ability and motivation of its staff to support this innovation?

Notes:	Description	Discussion	Enablers / obstacles
• What activities were used to develop the ability and motivation of staff in order to support this innovation? • What activities were carried out to support this innovation? • How were these activities carried out? • Who carried out these activities?	 Informal ☐☐☐☐☐ Formal		

Q3: How did structures and processes within your firm support this innovation?

Notes:	Description	Discussion	Enablers / obstacles
• What structures and processes were used to support this innovation? • What activities were carried out to support this innovation? • How were these activities carried out? • Who carried out these activities?	 Informal ☐☐☐☐☐ Formal		

Q4: How was this innovation supported by knowledge management activity?

Notes:	Description	Discussion	Enablers / obstacles
• What knowledge management activity was used to support this innovation? • What activities were carried out to support this innovation? • How were these activities carried out? • Who carried out these activities?	 Informal ☐☐☐☐☐ Formal		

Confidential

| Ref: | |
| Date: | |

4. Innovation performance measurement/indicators

Q1: What were the impacts from this innovation?

Notes:	Description	Discussion	Implied enablers / obstacles
• What were the expected/unexpected positive impacts from this innovation? • What were the expected/unexpected negative impacts from this innovation?	 Tacit ☐☐☐☐☐ Explicit		

Q2: How did you measure this innovation performance?

Notes:	Description	Discussion	Implied enablers / obstacles
• What measurement/indicators (e.g. stakeholder attitudes, business results, etc.) were used to measure this innovation performance? • What activities were carried out to measure this innovation performance? • How were these activities carried out? • Who carried out these activities?	 Informal ☐☐☐☐☐ Formal		

Q3: How do you intend to further develop/exploit the benefits from this innovation?

Notes:	Description	Discussion	Implied enablers / obstacles
• What activities were carried out to develop/exploit the benefits from this innovation? • How were these activities carried out? • Who carried out these activities?	 Informal ☐☐☐☐☐ Formal		

Confidential	Ref:	
	Date:	

Thanks for your input and co-operation. Transcript will be sent to you to confirm that I have understood what you have said correctly.

Please use the box below for any comments you wish to make.

Additional information

Appendix E: Company General Finding Report

1.0 Introduction

The aim of this report is to give feedback from five interviews highlighting key issues and suggesting potential, high leverage, 'quick win' areas for improvement in the innovation performance of ArchSME.

The findings are based on interviews which were carried out with one architectural technician, one architect, one project architect, one business development manager and one associate director.

The structure of the report will be structured around the following questions:
- What are the immediate innovations which ArchSME should progress?
- What is the current position?
- What are the potential problems?
- Why manage knowledge?
- What are potential improvement areas to sustain current growth?

2.0 What is the current position?

✠ Good at 'external' innovation to solve 'one-off' client problems.

BUT

✠ Not so good at 'internal' innovation to improve operational efficiency.
This finding is further explored and supported in the following sections.

3.0 What are the potential problems?
3.1 What is ArchSME's position?

Financial success Good at 'ring-fenced' team work	*Financially we are probably doing relatively well.* *All jobs are supervised by senior management talking to the people. We are work in a quite close team.* *Everything we are all in the team. That is the process and the structure – people involved.* *For something to be supported it, it needs to be shared . . . we share with the team, the whole team discuss it.* *To enable the relationship . . . it's more about the team building social event.*
Committed to architectural quality	*The way that I would judge the success is purely in achieving commercial success, but also achieving architecture success.* *If we judge on money, then we just do as well as everyone and not care about the subject. But we care about the passion by the designer and architecture.*
The firm is very young	*We are quite a young firm . . . . I have to . . . ensure that I am passing on my knowledge to younger members of staff.* *A lot of younger, less experienced members of staff, get a quite lot of responsibility.*

3.2 What are the potential problems?

Too busy, lots of work	*No one had time . . . we're too busy.* *Balancing sometimes. Amount of work we do within the teams . . . Sometimes, the work is too much.* *Time, we need time.*
Everything is done in 'ring-fenced' team	*[Different teams] are supposed to just wander around the office and comment on schemes, but they never have time to do that.*
Good ideas are not captured and further developed because of the pressure of the work	Organisation level *Some processes which have been set out, and then because of the pressures of work, just slipped.* *The pressures of work removed our ability to handle these sessions.* Project level *If we did do [assessing the project], then it will save time in the future and money from repeating mistakes.* *We should assess at the end of each project within the team. We should assess what went wrong and why, and we don't do it.*
Lack of appropriate structure and communication channels to encourage and support knowledge transfer between 'ring-fenced' teams and projects and in a formal way (e.g. post project review)	*We do encourage the communication between the team 1 and team 2 to share the information, but it is not always possible.* *Trying to increase our tacit knowledge throughout the company because we have a big problem with communication.*

NOT A PROBLEM NOW!!!

 **BUT**

With increasing growth of the firm, the limitation of the internal systems will probably become a significant restraining force.

4.0 Why manage knowledge?
4.1 What is knowledge?

Knowledge is largely tacit and is gained and refined through all activities, relationships (colleagues, clients and suppliers), experiences and observation.

Knowledge is knowing your role . . . knowing your place in the team to be gained.
Knowledge means the ability to carry out your job.
Knowledge is gained from experience from previous clients.
Knowledge is introduced and then must be shared for you train others to gain knowledge.
Knowledge means . . . what you've learn personally or tacitally from someone else, passed on knowledge,
[Knowledge] means our experience.

4.2 Where is knowledge?

Knowledge is mostly stored in heads of people.

The information source is the people rather than our client; rather than our product; not document.

4.3 What is knowledge management?

Knowledge management is more about 'people networks', not 'computer networks'.

Sometimes the admin team will come round and explain what they are going intending to do. It's by just talking to people . . . that's how information is collected in the practice.

4.4 Why manage knowledge?

The firm is in the creative and innovative business.	*We are in the creative business . . . we got the creative idea in the way we do things.* *What we do is design new technology.* *Most jobs are site specific any way.* *Quite often we try out new building components, materials, new products that we haven't used it before.* *I believe all I design is somewhere is innovated.* *Everything we design should be new, should be an idea to present, to develop.*
Knowledge is often shared and created when new situations are presented (e.g. new project comes in).	*The team meetings . . . The only thing that encourages knowledge sharing.* *Our industry is based on training . . .* *There is a process to sharing knowledge.* *Learning by doing. I am learning from others who have experience. That's the key within the practice.*
When employees require knowledge, trying to find the person they know rather than the right person to ask may be the only way of getting to the answer they need.	*We always share our knowledge if someone requires it.* *We need to close relationship between our colleagues within the practice, and also senior management and lower levels of staff to encourage . . . to seek advice when we need it.*
People prefer to receive information face-to-face rather than through on paper or electronically.	*Employees find more out from the informal discussion within the company really. I think it is quite rare that employees would have to look at the website to find something about the company rather than ask someone sitting next to you instead.*
Specification design in the past based on guesswork or trial and error.	*There have some new materials/new products we used that we haven't known enough about it, detail correctly.* *It's generally a sales problem . . . . Because it didn't provide enough information about products.*

4.5 What are the potential benefits of managing knowledge?

People spend less time in researching/accessing required information/knowledge.	*Our company structure, that's the sharing being able to share information and grow.*
People learn right across the organisation, build on mistakes and celebrate achievements.	*If the product isn't working . . . We can learn more about how the detail can be done correctly next time etc.*
Individuals and teams are links across remote locations or linked by information networks and communications mechanisms.	*It is not about the people in the individual office. They don't see other people during the day. The different groups interact at a social level. Through meetings informal, from gathering, social gathering. That was mainly.*
Improved sharing of information encourages better quality working relationships.	*You can find out more information if those suppliers are trusted.*
Product development cycles (e.g. drawing package) accelerate due to the availability and use of shared knowledge and expertise.	*We have people coming in from college. They have an understanding of design skills . . . they teach me.*
There is greater innovation and building on ideas of others.	*During sharing knowledge with my colleague, so I got this idea that we have this new material.*
Individuals are encouraged to develop and to grow their shared expertise.	*We encourage [employees] to develop themselves . . . we invest in them with time and money. We send them on training courses, pay for them to do courses on the web and also hold in house seminars.*

5.0 What are potential improvement areas to sustain current growth?
5.1 Immediate wins

Ensure there is a mechanism for capturing the outputs and new learning opportunities from future projects when they are completed	Establish post-project review policy, guidelines and checklists
Conduct 'exit interviews' when people leave, to capture knowledge which will be missed	Establish exit planning policy, guidelines and checklists

5.2 Short-term wins

Establish more formal structure system to capture and access knowledge context	Establish 'road map' to find knowledge in the firm Transfer people/knowledge between projects/business to spread and gain knowledge (particularly managers) (such as assignment system) Invest more in knowledge transfer (e.g. project briefing) rather than skill building (e.g. Learndirect project)
Create knowledge base	Establish 'products/components/ materials' database
Establish evaluation and reward system	Supplier performance evaluation (e.g. information accuracy) Link rewards to knowledge contribution and use through such means as the appraisal system

5.3 Mid- to long-term wins

Develop a knowledge management strategy	Role of IT (e.g. communication media) Innovation Competitive advantage Knowledge mapping
Link to human resource strategy	Align knowledge management strategy and human resource strategy

6.0 Summary: What are the key findings?

Current position		Good at 'external' innovation to solve 'one-off' client problems, BUT not so good at 'internal' innovation to improve operational efficiency
Potential problems		Not a problem now, BUT with increasing growth of the firm the limitation of the internal systems will probably become a significant restraining force
Potential improvement areas to sustain current growth	Immediate	Establish post-project review policy, guidelines, and checklists Establish exit planning policy, guidelines, and checklists
	Short term	Establish more formal structure system to capture and access knowledge context Create knowledge base Establish evaluation and reward system
	Mid to long term	Develop a knowledge management strategy Link to human resource strategy

Index